T0295260

Evolution

This book reconceptualizes the ancient philosophy of 'dualism' and a 'trinity' applied to classical and quantum nonequilibrium phenomena. In addition to classical mechanics and electrodynamics, a remarkable connection of this philosophy with quantum mechanics is established, which can be useful for quantum computing and the development of quantum artificial intelligence. Packed with the recent theoretical models, quantum simulations of black holes, and experimental observations of quantum phase transitions, this book brings a holistic approach that can be useful in refining the concept of the 'Creation', i.e., the evolution of the Universe from the condensed state of matter, and in explaining the artificial vision. This approach is:

- **Unique:** Connects ancient philosophy, classical and quantum mechanics, materials, and the latest quantum technologies.
- **Novelty:** High level of scientific interpretation of ancient philosophy, creation, interactions, and disorder.
- **Multidisciplinary:** Guides students to learn quantum simulations and develop philosophical ideas.

The book contains over ten years of in-depth research by the author on ancient eastern (Indian) philosophy and the latest scientific research in condensed matter and quantum technology. This book will simplify the picture of ancient Eastern philosophy, which can be read by the general audience, particularly undergraduate/postgraduate students interested in quantum information science without any mathematical treatment. However, some parts of the book will be ideal for senior academicians and professional researchers in both worlds.

Evolution

Classical Philosophy Meets Quantum Science

Somnath Bhattacharyya

CRC Press
Taylor & Francis Group
Boca Raton London New York

CRC Press is an imprint of the
Taylor & Francis Group, an **informa** business

Designed cover image: © Shutterstock

First edition published 2024
by CRC Press
6000 Broken Sound Parkway NW, Suite 300, Boca Raton, FL 33487-2742

and by CRC Press
4 Park Square, Milton Park, Abingdon, Oxon, OX14 4RN

CRC Press is an imprint of Taylor & Francis Group, LLC

Library of Congress Cataloging-in-Publication Data
Names: Bhattacharyya, Somnath, author.
Title: Evolution : classical philosophy meets quantum science / Somnath Bhattacharyya.
Description: First edition. | Boca Raton, FL : CRC Press, 2024. |
Includes bibliographical references and index.
Identifiers: LCCN 2023006780 (print) | LCCN 2023006781 (ebook) |
ISBN 9781032301181 (hardback) | ISBN 9781032303819 (paperback) |
ISBN 9781003304814 (ebook)
Subjects: LCSH: Creation. | Cosmology. | Religion and science. |
Quantum theory–Philosophy.
Classification: LCC BL227 .B525 2024 (print) |
LCC BL227 (ebook) | DDC 213–dc23/eng/20230522
LC record available at https://lccn.loc.gov/2023006780
LC ebook record available at https://lccn.loc.gov/2023006781

ISBN: 9781032301181 (hbk)
ISBN: 9781032303819 (pbk)
ISBN: 9781003304814 (ebk)

DOI: 10.1201/9781003304814

Typeset in Palatino
by Newgen Publishing UK

Contents

Preface

From the available historical artifacts and mythological stories, it is generally believed that ancient civilizations, particularly the Sumerian, Egyptian, and Indian civilizations, maintained very good connections, which helped to conceptualize a model of the 'creation' of the universe. The pre-Socratic-era philosophies show many similarities with the older wisdom literature and mythological cosmogonies of the East, particularly India. Based on the cycles found in nature and the observation of heavenly bodies, an ancient scripture the *Rig Veda* was compiled before 1500 BCE. To interpret this huge text, a few schools of philosophy emerged, and the concept of the smallest indivisible entity, the 'atom or voids' was put forth. In search of energy within atoms, a philosophy of the number system called 'Sankhya' philosophy was developed around 1000 BCE describing the evolution of nature without an active creator. The central idea of Sankhya philosophy is the 'dualistic nature', which is termed as uncertainty or fluctuations. Other aspects of Sankhya philosophy, namely three elements of opposite nature creating a state of equilibrium, can be found in the works of the great post-Socratic philosophers that laid the foundation of Western philosophy as well as science. Going through the era of classical mechanics and electrodynamics, we shall discuss a remarkable connection of this philosophy with quantum mechanics, which is useful for the emerging field of quantum computing. In addition to the recent theoretical models and experimental observations of matter, Sankhya philosophy can be useful in refining the concept of creation and evolution of the Universe.

Preamble: Philosophy through Careful Observation and Self-Realization—The Birth of Cycles
Chapter 1: Creation? A Cosmic Model—Birth of Matter and Mind: Two Cycles
Chapter 2: Background/History—Understanding Creation: Gods and Goddesses
Chapter 3: Life Begins on Earth: Creation of the Devices, Deities, and the Concept of God: Creation of Cycles
Chapter 4: Theory of Evolution—(Sankhya) Philosophy of Dualism: Creation of a Cycle
Chapter 5: Interactions at the Macroscale—Classical Mechanics—Equilibrium and Rotation: One Cycle
Chapter 6: Energy/Light and Space (Matter) Application in Condensed Matter Physics: Two Zeros
Chapter 7: Condensed Matter Physics—Creation of Particles—Bosonic vs Fermionic: One and a Half Cycle
Chapter 8: Creation through Interactions at the Microscale—Quantum Mechanics
Chapter 9: Space–Time Distortion: Black Holes, Condensates, and Quantum Devices
Chapter 10: Quantum Simulation, Artificial Intelligence, and Vision
Conclusion
References: Suggested Literature
Index

Acknowledgment

Dedicated to Sri Ramakrishna Paramahamsa

My immense interest in understanding the primordial energy (Goddess Kali) and the origin of rotational motion propelled me to develop the concepts given in this book. I would like to express my gratitude to God (Divinity), who has guided me to develop the ideas that I present in this book. The main inspiration for the book came from my study on Sri Ramakrishna Paramahamsa. In 2017 and 2018. I had the opportunity from Ramakrishna Mission Cultural Institute, Golpark, Kolkata, India, to present talks that initiated this work. In 2019, I presented a seminar in the School of Philosophy in my university (University of the Witwatersrand, Johannesburg). In 2020, I received an offer from Taylor & Francis to write this book. However, I spent more than ten years studying ancient scriptures and developing philosophical ideas. In parallel, I performed both experimental and theoretical studies ranging from quantum electronics to quantum computation, which helped me enormously to compare the scientific results and fundamental science.

The main parts of the book were compiled during my sabbatical leave in 2014 (University of Leeds, UK) and 2019 (MiSIS, Moscow, Russia). I am thankful to E. Flahaut, University of Paul Sabatier, Toulouse and Ravi Silva, University of Surrey, Guildford for inviting me to their institutes which was very helpful to complete the book. The University of the Witwatersrand in Johannesburg, South Africa, where I labored on this book, has my sincere appreciation. My students and colleagues deserve praise for their assistance.

This philosophical part of the book is inspired by a few books namely, 'Kapila's Sankhya Darshan' by Sri Panchanan Tarkaratna, and `The Philosophy of Gita and Sri Ramakrishna Kathamrita' by Harishchandra Sinha. The late Sri Krinajiban Bhattacharyya (Bhatpara, WB), who mentored me in the development of my philosophical thoughts, has my gratitude. Late Sri Satyanarayan Mukherjee (Agarpara, WB) provided guidance and stimulating discussions. I am incredibly blessed to have grown up in a loving family and community (Prakriti). My father (Hara Prasad Bhattacharyya), mother (Anjali Bhattacharyya), grandmother (Ranibala Bhattacharyya), and aunts (Barama, Pishima, and Jethima) had an influential part to play in my early intellectual and spiritual development through their example and teachings.

Through my spiritual experience with God, I felt that I have the power to change my future if I devoted myself to the Goddess Kali. There have been times that I have been inspired as I have explained concepts to those who have listened to me. As a child, I was inspired by nature as I observed a huge field that almost touched the horizon outside my schoolroom window. This experience taught me about space and infinity as I became absorbed in nature as a child. Naturally, I am a follower of Einstein from school life.

I am very thankful to a young scholar Mr. Samya Mukherjee (Kalyani, India) for supplying me with plenty of information on science and philosophy, which acts as an introduction to my writings. He contributed to the slokas, mythological stories, and scientific terminologies. I have used some of the calculations and data analysis of Mr. Shaman Bhattacharyya in the book, who not only transcribed most of the parts of my book from my diary but also critically revised the work. Subhas Pal is acknowledged for some of the artwork.

My friends Neil Coville (Wits University, Johannesburg) and. Angela Simon (Vacutec, Johannesburg) very carefully read the manuscript and made suggestions about the content of this book. Stimulating discussions with them were very useful in improving the presentation. I acknowledge Joy Raychaudhuri for his contribution to the introduction of the book and Shanta Bhattacharyya for proofreading.

A special thanks should be given to Mr. Sibsundar Samanta (Bhatpara Sahitya Mandir), who inspired me over time and provided some of the most profound books used as reference here. I received substantial inspiration from my former students and friends, particularly from Christopher Coleman. My former students Christopher, Davie, Dmitry, Declan, and Shaman produced the results used in the book.. I received inspirations from friends Victor, Svetlana, John Mathieson, Dibakar, and Kalyan Bhattacharyya.

A special credit should go to the publisher Dr. Gagandeep Singh, who invited me to write the book. Fred Dahl and Indhumathi Kuppusamy is acknowledged for editing the manuscript and other production-related work. Thanks to all reviewers for carefully reviewing the synopsis and proposal of the book and providing some of the most valuable suggestions. Preecha Yupapin gave valuable suggestions to improve the content of the book, and I use his research work in this book.

Finally, I must acknowledge my wife, Mrs. Monali Bhattacharyya, for her ongoing support of my work.

A memory (that cannot be erased) on an equation that applies everywhere. The Holy Mother reminds you 'I am available'. This can come to you during sleep like a dream, but it will not leave you when you wake up but remains permanently with you. In daily life, we face many problems and tolerate bad interactions and forget most things as a part of our volatile memory.

Let us recycle our bad memories into good ones and finally into one.

Author Biography

Professor Somnath Bhattacharyya, PhD
Head of Nano-Scale Transport Physics Laboratory, School of Physics
University of the Witwatersrand, Johannesburg, South Africa.

Biography
Somnath Bhattacharyya is a Professor in the School of Physics at the University of the Witwatersrand, Johannesburg, South Africa. His research is focused on the area of condensed matter physics, nanotechnology, and quantum computation. In 1997 Somnath Bhattacharyya completed his doctoral degree from the Indian Institute of Science, Bangalore in Condensed Matter Physics. He worked as a researcher in the USA, Germany, France, and England. His major interest is in the transport properties of carbon and his major achievements include multilayer carbon-based resonant tunnel devices, high-speed transport and spintronics in nanostructured carbon devices, n-type doping of nanocrystalline diamond films, spin-triplet superconductivity in boron-doped diamond films, theoretical models for transport in disordered carbon, and recently, quantum simulations of many body systems. Prof. Bhattacharyya has published two books, four book chapters, and over one hundred papers in peer-reviewed journals. At present Prof. Bhattacharyya is engaged in developing a new infrastructure for a wider range of nanotechnology that will include quantum bits, quantum matter, and also quantum simulators based on hybrid carbon structures.

Somnath's studies, however, are not limited to just science; he has spent much time investigating one of the most abstract and elusive and yet fundamental subjects, the concept of god, as w ell as answering some of the most paramount yet enduring questions about reality. Through his research, Somnath has constructed models linking such scientific theories with ancient concepts whose origins reach back many centuries. Through examining a variety of artifacts displayed at various locations around the world and numerous different religious and philosophical texts, in addition to several works dealing with metaphysics and mysticism, Somnath is uncovering invaluable information connecting and explaining some of the world's most influential religions, philosophies, and ideas in very profound ways. His main interest lies in the understanding of creation. Somnath is the co-founder of the Society for Agni Vigyan Advancement (www.agnivigyan.com/) where he bridges the gap between science and philosophy. Through the display of his photos, he tries to simplify some of the complicated concepts of philosophy and science.

Background/Introduction

Philosophy through Careful Observation and Self-Realization

In 2019 around December during Christmas time, with the year coming to an end while dreaming of an easily accessible quantum computer and artificial-intelligence- (AI-) controlled world in 2020, we wouldn't have anticipated the lockdown and the virus. But we were then struck suddenly by an impending pandemic, and by early summer 2020 the ambulance sirens were the only sounds that wailed in the deserted streets. In a matter of a few months, the whole world was seen struggling with financial losses, unemployment, and a fragile health care system. Our egos are shattered; people were running recklessly in the streets of New York, Mumbai, or Paris with the hope that this too will pass. A virus had run amok, and we were struggling to keep it under control. The looming uncertainty about the future was what brings us to the core topic of understanding what we call 'intelligence' and 'consciousness' with overlapping views from modern science and spirituality. Over the last 3,500 years in ancient India, philosophers and spiritual practitioners, scientists, and academicians have been debating this topic and putting forth their insights; this was also the subject of the standalone dialogues in Plato's cave allegory in the *Republic* and Aristotle's *Discourses*. It would be difficult to summarize the whole development of Eastern/Western metaphysics and philosophy, but, here, we would like to focus on one insight, which is the Darshan or Vision in the Western world called 'philosophy'. The ancient Sankha Dwaita (dualism) Darshan and (nondualistic) Adwaita Vedanta concept of consciousness and matter would be useful to explore the recent developments in consciousness studies in modern cosmology, condensed matter physics, quantum many-body physics, and quantum biology and neuroscience. Here, we are not trying to find the unified law of energy, which is a dream of all physicists; rather, we try to understand the interactions process. The microscopic quantum objects, which should be treated as the spin centers and the dynamics of the force field out of the equilibrium state, will be studied with the help of a quantum processor (computer). Understanding the whole process of evolution, including creation, preservation, and dissolution, will control growth. Condensed matter and many-body physics have been established to understand the interactions between the spins that may be applied to microscopic objects where the presence of life has remained debatable.

To start this exploration, we must begin with a set of questions to identify what we want to explore! Questions are important because they formulate our doubts, or, rather, they substantiate them with clarity. Let's say you are looking at the painting by the Frenchpost-impressionist artist Paul Gauguin, 'Where Do We Come From? What Are We? Where Are We Going?' These are the same set of questions that were in the minds of the visionary seers, or Rishis, of ancient India some 5,000 years ago. What's my relationship with the universe or the world outside? And is there some supreme intelligence or God that made all these possible? The questions are deeply connected with the nature of our experiences, with the phenomenal world of our experiences. We can improve/modify the questions in the 21st century. How do we explain the interactions at the microscopic level, which are associated with the uncertainty of the states? How do we control the growth of tiny objects (like continuously breaking symmetry) that are evil to society or life? What is the future of our civilization, either its natural (spiritual) life or its artificially created objects? Why does uncertainty or duality work? Why is there disorder, and what is the nature of true disorder? Which side of the duality is true? Ancient texts such as Seer Kapila's Sankhya philosophy presented the concept of duality; however, is there any experimental proof? How can we verify ancient philosophical concepts?

The philosophical idea of nature, the universe, and its creation began with the compilation of the first volume *Veda*, i.e., *Rigveda* in India more than 3,500 years ago based on the movement and relative position of the star constellations. Starting from the worship of the Sun God (in different forms), it described a universal picture of cosmogony with dynamics and a time-dependent process of how equilibrium can be attained from an apparent nonequilibrium state. Not only physical processes but also biological processes in the human body and mind have been described based on the law consisting of three elements having entirely different characters. It also describes how many Gods can be united to a unique, singular God and His consort (a dualist concept). *Veda* addresses Agni (or fire), which creates intelligence in our brain i.e., creation, which evolves into sensing organs and activities. Today we are trying to develop more advanced artificial intelligence (AI) for devices or machines based on a quantum neural network that can be related to an ancient mathematics and number system, which is described in the proposed book.

It is noted that *Veda* or *Rigveda* is very vast, and common people can't read it easily since the philosophy part is hidden. *Rigveda* was reinterpreted by great Indian philosophers, namely Seer Kanada (Vaishasika philosophy), Seer Gautama (Naya philosophy), and Seer Kapila (Sankhya philosophy). The model of evolution is popularly known as the 'Sankhya philosophy', which deals with the elementary mode of vibration and matter at the smallest scale, which is known today as quantum fluctuations and elementary particles. This metaphysics describes the transition from a closed (zero) to an open (nonzero) quantum system or from a quantum to a semiclassical system and tries to solve problems of humans such as mind, intelligence, disorder, and sensation. In this modern approach of philosophy, the idea of a creator is removed, and then a simple picture of creation based on interactions, which looks like a law of creation and is open to anyone (not based on a religious concept), is introduced to give a holistic explanation of the structure of the universe. It is developed as a number system based on various astronomical observations; however, it is applied to biological systems particularly to understand the brain-to-body relationship and sensations. It has influenced classic mechanics and geometry over several centuries and more importantly neuroscience. Most of the ideas developed around this philosophy are in the 19th century, i.e., before the birth of quantum mechanics and the theory of relativity. Recently people started comparing ancient philosophy with cosmology since there are many unanswered/open questions in physics. In this book, I shall describe its application in classical mechanics and then to a quantum system that is important for present and future quantum information technology.

Creation starts with small-scale objects and small oscillations, which eventually emerge as a large object. This is like the manifestation of a nonmanifested entity or a quantum to a classical state. In this book, the laws of classic mechanics, electromagnetism, and quantum mechanics are explained with the help of Sankhya philosophy without any mathematical treatment. Quantum computing system works based on the transition of a two-level system driven by waves, which are described as creation, retention, and destruction. This process can be made visible when many atoms contribute in an entangled manner like a neural network. So it is like the transformation of knowledge or transformed into work or senses through processing or intelligence. The complete process described in the philosophy of Sankhya is explained in the book.

Thanks to the distinguished authors C. Sagan, F. Capra, R. Dawkins, M. Kaku, and others who already showed us the direction of scientific advancement and the future of civilization. But we need mathematical expressions that can describe the interactions on the microscopic scale. Instead of parallels between the modern science of the macro world and Eastern philosophy, a recipe for the microscopic world is required to clean up the polluted environment. We must find precise applications of our knowledge and experience through a transformation (Rajas). Let consciousness (Satwa) remove all superstitious beliefs (Tamas) and open our eyes.

> ॐ भूर्भुवः स्वः तत्सवितुर्वरेण्यंभर्गो देवस्यधीमहि धियोयो नः प्रचोदयात् Rigveda 3.62.10
>
> *Oṃ bhūr bhuvaḥ svaḥ tat savitur vareṇyaṃ bhargo devasya dhīmahi dhiyo yo naḥ pracodayāt–*
>
> [Om, Earth, Space, and Heaven; and Lord beyond these who is Reality (Truth), Consciousness and Bliss. We meditate upon the glory of that Supreme Being Savita (Sun as the symbol of God) who has created this Universe; who is fit to be worshipped; who is the embodiment of knowledge and light; and who is the remover of sins and ignorance. May he guide and illuminate our actions and intellects at all-time in the right direction.]

The *Gayatri Mantra* from the ancient scripture *Rigveda* addressed the Sun who lives in three worlds as fire. Time on our planet started with the birth of the Sun followed by a transformation process in the planet that created Life. Sun cleans up our polluted atmosphere through a cyclic transformation. However, we would like to know about the situation before the Sun (or bright stars) were formed and how the transformation takes place in the Sun that provides us energy.

In this work, I will attempt to overlay creation myth and philosophy with science. This is will not be an easy journey; it will be difficult to understand for many. This vision has not been an easy one for me to convey. If you persevere in attempting to understand the principles outlined in this work, I hope to bring to you a greater understanding of creation and in so doing give you a greater understanding of the present, the world, and who you are. This has been a difficult journey for me as this vision has unfolded to me, and I have tried to use words to convey the ideas in this vision.

It took me a long time to reveal a universal model of the creator or God, and I am still improving my thoughts. Initially, there were many gods, which are thought to have originated from one unique god. However, the unique god can take many forms or phases to perform different kinds of work. One God lives at the center of a complicated world made of many layers. This reminds me of a multiverse or a metaverse as the science of films is presenting today. In this introductory chapter, we have explained different stays of creation, preservation, and dissolution. Also, we have seen a conversion from energy to space to sound to mind and finally the re-creation of energy at the end of the cycle. We have explained Atman, Paramatman, and Maya-Shakti, which represent different stages of creation. We explained the cycle of creation with the duality and trinity models and compared it with nature and the planetary system.

Preamble: Philosophy through Careful Observation and Self-Realization—The Birth of Cycles

Energy

I started to imagine that, as the creator (Brahma), I was present and resting on the bed of water for a long time in a state where energy and mass or light and darkness were inseparable. I was in a dormant (Susupti) state filled with happiness, although unhappiness (some disturbances) was about to grow around me. However, I wished to bring some change to my life. Although I did not make any large movement, I was able to make conformal changes that include splitting my body. To express my satisfaction, I wished to create my mouth and open it to produce deep and resonating sound. Then I wished to create my vision and opened my eyes. I wished to change the shape of my body through rotation that would create voids and release a part of my body or my consort (i.e., Maya Shakti) like heat energy and Agni (Fire). Although I wished to do these activities, I was experiencing resistance. I was in a state of dilemma and thinking about how to overcome it.

Space

I extended my wish to overcome my world of imagination (dream). Finally, I woke up, sat on the water bed, and started swinging my legs. I didn't have any other activities except for oscillating periodically, which meant creating a superposition of states from two ends of the cycle. Thus I split my legs into two parts. Then I decided to make nonharmonic oscillations and then expansions. Initially, the swinging was periodic and then became nonharmonic due to absentmindedness. I broke myself into two unequal parts and then tried to communicate between these two pieces to balance them. In that case, each leg was divided into two. Initially, the motion was circular and had concentric orbits. Once these things happened as a result of carelessness, nonharmonic oscillations start. After dividing myself, I found many followers who also divided themselves. A chain reaction started around me. The two parts are energy and matter, covered by a closed shell. Then two shells were created, each containing two parts. These two shells were covered by a larger shell, and a second large shell was formed.

The dualistic view of the universe can be described with this super-shell model, like eggs containing two yolks or central parts of opposite natures, which are souls Atman and the greatest soul Paramatman. The shell is called Maya or Shakti, the primordial energy. I am the Paramatman surrounded by many Atmans who will follow my work or activities.

DOI: 10.1201/9781003304814-1

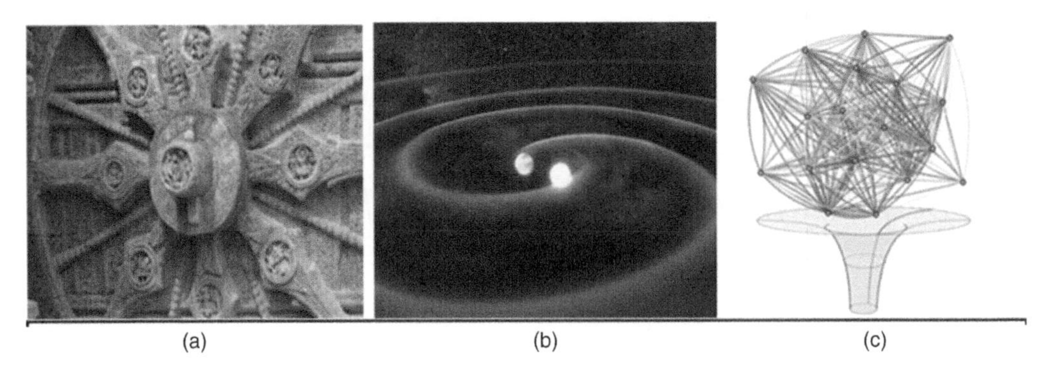

FIGURE 0.1
Artifacts (a) the Sun and (b) the wheel used to develop scientific models of the universe and condensed matter physics. A large wheel carved out of stone in an ancient Indian chariot-like temple. The thick axle of the wheel also has a symbol carved into it. (b) A binary star system is creating gravitational waves. Two intertwined spirals, each spiral starts from a star-like bright object. (c) A model using the folding of space is used to describe a maximally chaotic system.

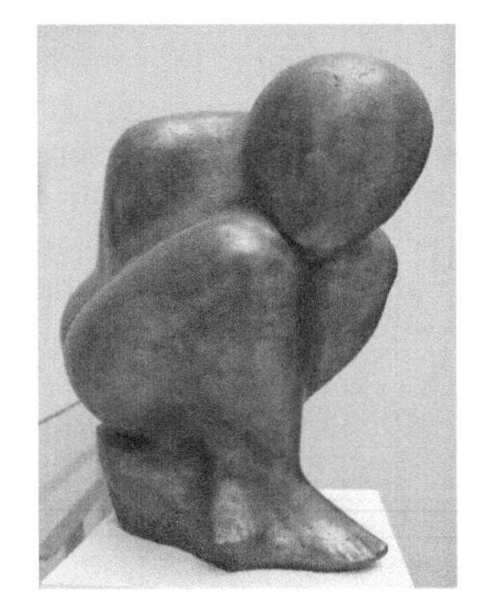

FIGURE 0.2
Before creation, there was no space or time. At the time of creation a globular shape of the body was developed. It looked as though several ellipsoids are coupled together before creation.

Although I had been One (lone), I always had the idea of playing with two states, or duality by the superposition of two energy states. I wanted to create a mixture of these two without losing either of them. I split myself into two parts like creating two eyes or two legs. I can create my subtle body (Lingam, having a positive curvature or outward flow), which is like a funnel-shaped vortex in space having a strong attractive field, and look for another body (Yoni having a negative curvature or an inward flow), which has opposite properties of the vortex. So from my neutral nature of the whole universe, I created a polarity of local structures like the Sun. From one Agni or Sun, I created many centers of fire just as the Sun creates fire on Earth, who will follow my style. I created a variety of space (matter) having a wide range of energies (speed) that are connected with me through an attractive force (Love). However, I love my creatures equally. My children extend their love to one another continuously depending on their potential and by maintaining some distance. I watch my creation all the time.

Vision

I looked at the night sky, which is filled with thousands of stars. I felt as though a big person with a thousand eyes was looking at me constantly. A man with thousand eyes has a thousand heads and hands and legs etc. Out of this large number I found that two were very close to me: in the day, a ball of the fire (i.e., the Sun) and in the night, a shiny ball (i.e., Moon). Both of them gave me company as if they complemented each other and are therefore considered to be entangled. I see a superposition of the Sun and the Moon as the source of light and life on Earth. So the duality model still worked in my mind. I would like to depict the model of the Man with a thousand eyes as my head, two eyes as the Sun and the Moon, and four sides as four hands instead of only two physical hands. I could imagine myself as a big bird with two large wings representing the Sun and the Moon with my head corresponding to the sky or space. I stood on the surface of the Earth, which is not flat but appears to be an egg. Hence, a bird egg model was developed. With my consort, I could look after the planet in a better way. So I split myself into two birds who concentrate on the egg.

After creating the fire, I looked at the shape of the flame. I needed to understand how to visualize the creator (Brahma) or heat in the form of an egg of the creator (Brahmanda). A stable flame looks like a luminous egg, almost an ellipse or an ellipsoid that is also pointing upward. But the inner part, particularly the center of the flame, appears to me bluish. So the flame is red, and the burning of the wood is black or blue like a golden egg with a dark yolk. However, at the boundary of the flame, a different color is seen, which protects the flame from external disturbances. In the flame, a white part can also be seen. Now the fire is split into multiple flames, and two neighboring flames can interact and enhance the intensity of their flame parts. The central dark part of the flame is the interface of the matter of the fuel and the nonmatter of the plasma of fire. This observation inspired the cosmological model that all bright things can have a dark center. In the sky, a flash of lightning originates from the dark cloud. The origin is hidden in a black core that contains all reasons like a black hole. Finally, I look at creation on Earth.

FIGURE 0.3
Ancient model of creation as described in some scriptures. A Tibetan artwork shows the interaction of the Sun with the Earth. Universe: Union of Purusha (having many hands) and Prakriti. A Tibetan sculpture of a being with a boar and bull-like head and many arms radiating in various directions. He is embracing a woman.

I, the first man (Purusha), wanted to be dormant and invisible, but I watched the activities of others. I expanded as I became hungry. I became excited as I felt disturbances from the surroundings (such as extreme temperatures and their variations). The space around me was not only distorted but also ruptured, and I rise like lightning or a volcanic eruption. I am created when the duality is broken, i.e., from the interplay of three different and opposing characteristics.

Creation

The first (Woman) person, in the form of energy (Shakti), is manifested for the first time overcoming all disturbances and resistances. Fire is within us, inside any living creatures (as well as materials), which gives us life. I should protect this fire through survival in the form of a human body. Instead of a small body, now I have to deal with a **coarse body**. At first, I am looking for a way to survive in the cold weather. I start to rub two pieces of wood together that produce fire. To survive, we most importantly need space, then food, water, light, and sound. After acquiring space, I need a source of heat. That is when I need fire. Fire gives the light needed. However, my appearance is covered by the smoke and I cannot expand. Smoke resists my emergence, and I overcome this resistance by increasing my anger. Then I need a weapon to procure food. I will have to use controlled movements (periodic oscillations) to cause a high velocity so that I can hunt the animal. However, the animal also has to be targeted properly, so I invent the bow and arrow. This works by restoration force. The arrow is drawn back to a certain distance until I see that the string can exert enough force on the arrow to give it an acceleration so high that it hits the target and penetrates it. The operation of this device develops laws of motion and philosophical thoughts. This knowledge will lead to machines, inventions, and the wheel, which is the main element of machinery. The rotation of a wheel in action and its periodic manner will run civilization.

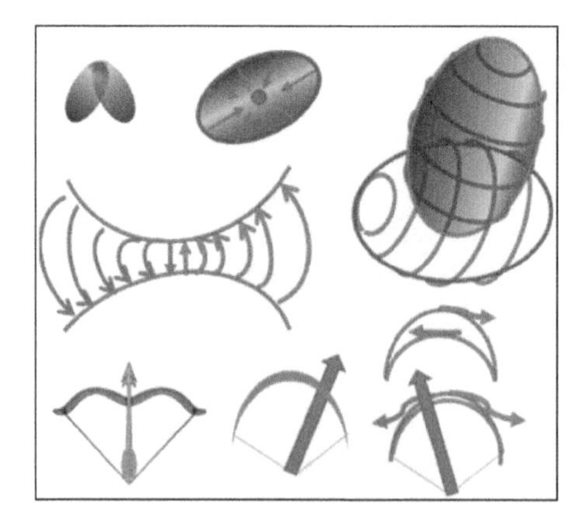

FIGURE 0.4
A model shows a division of space, a collapse of space, and the formation of an egg-like structure. Two ellipsoids are connected by tunneling in the curved space. It is comparable to the bow-arrow model. Starting from one we find two split ellipsoids which are also overlapping. Two curved lines connected through arrows. A bow is created formed from the curved line and an arrow is pointed in different directions.

Preservation

I create heat and light by rubbing my hands and swinging my feet repeatedly. Having created fire, I have to care about its sustainability. I need fuel to maintain my form, i.e., fire in the form of five requirements: materials (solid, liquid, and gas), heat/light, and space. Just like an open stove, the heat and light emitted by the Sun are useless unless they can be harnessed. This is done by using a cover. The smoke created by the fire can be managed in this way so that my temperature can be elevated. In the case of the stove, a pot is placed on the stove. In the case of the Sun, we need something that rotates and periodically absorbs the light and also blocks it. For this, we have the Moon. These three elements form a trinity. I can be present in these five forms as concentrated heat. So I develop five senses to reveal my form: sight, hearing, smell, taste, and touch. Fire can be utilized in five activities, or five types of mechanical movements mostly associated with the five organs and senses. Fire seen through a small opening shows its power inside an insulating body. This process of

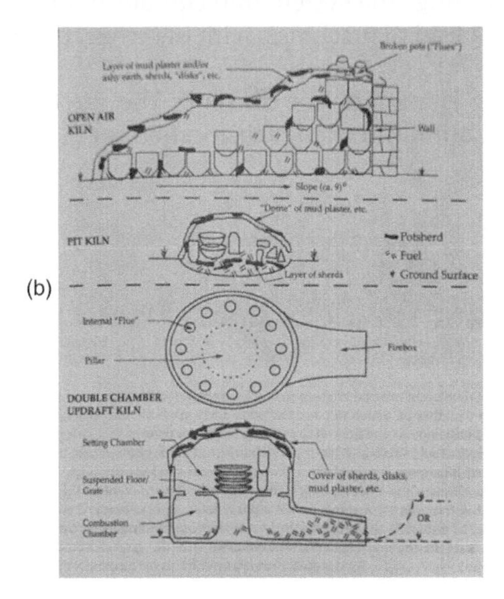

FIGURE 0.5 (a)
Relationship between the Sun and the Fire. Creation of a spectrum of colors. The temperature of the fire can be largely increased through confinement. The heat produced from the fire remains at the center of the furnace and is recycled through an egg-shaped or ellipsoidal chamber.

FIGURE 0.5 (b)
This structure was routinely used in ancient times and even today to sustain fire over a long time for smelting minerals.

holding the fire increases the temperature. This is 'Shakti', who also works in the human body. Potential energy can be increased by feedback, i.e., repeating the process. How, then, can one increase 'Shakti', or potential energy? It is like winding a string or spring (a serpent as a symbol), and coiling can be done if you have a dual nature (i.e., the head and tail of a serpent are the same). There is no beginning or end. Start with a straight object and bend it into the form of a closed loop. So two flames interact to increase the energy. It ultimately passes through a resonance, which corresponds to the highest temperature; a third eye is developed from two normal eyes. Beyond a dual state, this resonance is stable or at the level of equilibrium. I split myself into many parts like many flames, which can even look different in colors. In this process, I expand myself.

Dissolution

I live at a deep level of the Earth as a very compact form; however, when I come out, I try to expand rapidly. Through my power, I can consume everything as I move very fast. The fire is growing continuously. The flame is spreading out from its origin and tries to burn everything to create all of space. But I have to control the divergence and stabilize the reaction. I keep on striking every material that wants to stop my activities. Finally, I remain confined in a place where equilibrium is established between heat and cold (water) through a water cycle. One day, the fire leaves the body and goes to a place of higher altitude. Collecting all heat, a cloud forms in the sky, which blocks the light of the Sun. Absorbing the power of the Sun (also through the reflection from the Moon), the cloud in the sky becomes heated up. Ultimately, the cloud is dissociated into raindrops through lightning, so the heat or fire is converted into another form of energy. I overcome the attractive force of our planet and then use the attraction to complete the cycle. I neutralize the repulsive force addition of an attractive force that brings the system into equilibrium through the completion of a cycle. This knowledge will lead to machines, with the wheel being the main component of machines.

In the universe, nothing is immortal or constant. The fire I produce cannot be located continuously all the time, but it can change its location for better sustainability. My body

FIGURE 0.6
The Universe, consisting of three characteristics, can explain the state of equilibrium in mechanics. The central character is holding a ball of fire (the Sun) or an ellipsoid representing the universe. It is the egg of the creator (Brahmanda).

supported me for my ignition, but parts of me became old and unusable. The superposition of matter and energy is about to be lost. So I collect the fire from the bodies that cannot support the periodic motion since the motion becomes nonharmonic. One day, fire leaves the body and goes to a place of higher altitude. Collecting all the surrounding heat, a cloud is formed in the sky, blocking the light of the Sun. Absorbing the power of the Sun through the reflection from the Moon, the cloud heats up and becomes excited. Ultimately, the cloud is dissociated into raindrops through lightning, so the heat or fire is converted into another form of energy. In this way, I can overcome rebirth.

Sound

So far I know how to see through my eyes and move my body but cannot communicate. When I feel disturbed, I became extremely angry. I open up my mouth and express my anger. I break up space and matter into pieces. I become afraid of scary animals although they contained me. I have to shout at them to disperse them. I would be hungry and become a hunter. I can kill animals and therefore become lord of the animals. But I have to love them; I have to attract them by using my voice. Finally, I would like to express myself verbally to connect with others. I open my mouth and create space; like opening the mouth of a pipe, a sound like 'AW' is created. Instead of keeping my mouth wide open, I close it and create a small opening and the sound 'OU' is found. Finally, I try to make some 'Ma (or Mom)' or 'OM' sound without opening my mouth. A combination of all these three tunes produces all kinds of sounds. Sound is produced when we focus our mind or thought on a subject and release the energy or air. So all these processes are produced from the exchange of heat; a transformation of energy; from an attractive to a repulsive force. Now, I try to create my mind or fluctuations.

Mind (Prakriti)

A combination of attractive forces to particular point results in a repulsive force that fluctuates with time. Our mind has an attraction that tries to localize us through a binding force. Mind (called 'Chitt') is also fluctuating like a wave that gives false information (duality) to the observer. However, it can be focused on a particular point, which is well defined as the truth (called the 'Sat' or 'Satwa'). After passing this focal point, the mind diverges as if it has gained some repulsive forces. An example of this process is heat-induced transformation. When heat or light waves focus at a point, they can disintegrate the material and separate the parts. In mechanics, we concentrate on a target using an arrow and then release it. After the release, it produces happiness in the mind (or Ananda). Heat is a signature of joy or love that produces more vibrations between atoms or particles and effectively creates repulsion, which can overcome the dualistic attractive force. So the fixed body (*Sat*), attraction (Chitt), and joy (Ananda) complete the universe and are called 'Sat-Chit-Ananda'. In this holistic picture of the universe, we always have to overcome Maya, or the attractive force or duality, and enter into the regime of repulsion through a point of neutrality (Sat). A state overcoming the Chit can be compared to consciousness,

FIGURE 0.7
Fire lives in three forms in three worlds; the Sun lives in outer space, the lightning in middle space, and the wild fire on the Earth.

which many people think of as a field or as a force field that is extended everywhere. It is the cause of any movement or acceleration, which causes currents or flow. This is possible when two opposite characters are present, or it works between two of the same characters of different quantities. It is like the storage of energy that always has two opposite poles that are the fixed points to truth (Sat). Current (or Chit) or consciousness flows between the poles and produces the heat from the resistance that is always there. Without resistance, there is no relationship between the flow and the tension (or between acceleration and the force). This principle works throughout our life. These three elements form the universe even if they cannot be visualized or depicted. However, we shall try to understand this through models in the latter part of the discussion.

The re-creation of energy (heat) to the emission of heat through light and sound leads to an equilibrium (pleasure, joy). Finally, I created cycles and split them into halves. This concludes my dream and I wake up.

Creating Models

At the end of my vision, I try to develop a (scientific) model of the creation of matter and force field in the universe based on the thoughts of the ancient philosophers who searched for a holistic model of creation through the analysis of the human body, mind, and intelligence. While it is believed that something was always present, something that observes the whole process, the details of the transformations were hardly understood. Based on different scriptures and scientific models, we can think that, in the beginning, everything was dark, cold, and silent. Due to instability, a dualistic view of mass and energy can be imagined. This action results from the interaction of two parts in a curving or wrapping space. After rotating the two halves of opposite character in a spiral space, matter become dense, thereby increasing the pressure and leading to a collapse (space–time distortion). From the burst of the condensed matter, energy particles are emitted. The fragmented space is responsible for the creation of the universe. The outer surface of the space consists of rapidly fluctuating fields that cover the absolute static and supermassive interior: the seed of creation. I, the Brahman, as the cause of the creation process of energy, space, and time, have always been present. However, I created a view that can be dualistic (finding the contrast or the opposites) and extended myself to a higher dimension (a trinity).

Dualism

Initially the model was always based on the Sun and Moon, which are two points. Every day, the Sun rises in the sky and descends, which develops the model of the Soul's (Atman's) afterlife. How do we connect two points using a curved line between the initial point and the target? *This is the principle of constructing the bow and arrow (or a projectile).* You are supposed to launch the arrow at a particular velocity, and the angle concept can be extended for Atman after releasing the body as it should meet the Sun with a very high initial speed. With that practically, it reaches the cloud and contributes to rainfall. The Soul (Atman) with low speed cannot reach heaven or cloud and returns to Earth, unlike the other case that attains liberation. *This concept matches modern cosmology, which deals with an open and closed universe connected to the big bang or a steady-state universe model.* So the concept of duality can arise from these two points or two objects like the Sun and Moon that overlap, resulting in an eclipse. Again I must admit that this is a holistic model of the universe and that several enthusiastic scientists or philosophers have tried to compare these with observations or models in modern physics. However, no details, particularly mathematical treatments, have been given even though remarkable similarities have been observed. This is the motivation for the present look. The motion of the Soul (Atman) can be compared with the cosmological model involving the big bang or steady-state universe. Although the universe is made of matter and intelligence, like a seed, has two halves, can be well described as a necklace consisting of a flower bound in a string. While awakeness can be compared with the attractive gravitational force, the state of sleep without dream represents a great repulsive force.

Trinity: Three Opposite Elements

Three is one: one egg and two birds or one bird and two eggs, One Man (Purusha) and one Woman (Prakriti), in combination, form a circular orbit system. Sunrise and sunset

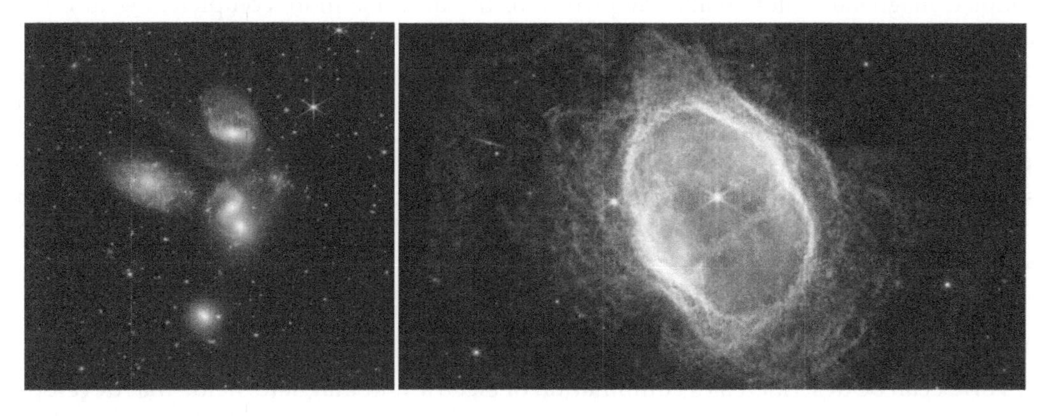

FIGURE 0.8
Ellipsoidal structures of galaxies and egg-like globular clusters of stars were observed by the James Webb space telescope in 2022.

suggest the rotation of the Earth. The change of phase of the Moon also shows daily rotation around a center. But the revolution part is missing, which adds another center, and the orbit becomes elliptical. One elliptical orbit holds both rotation and revolution of the celestial bodies. Two different rotations can be described as one being static and the other relatively dynamic. All motion is relative. The Sun as a static object can see the Earth and the Moon; one is static and one is dynamic. So we need two poles, one static and one dynamic. A supreme creator (Purushottam or the greatest Purusha) can hold both static and dynamic phases. Diagrammatically, He lives at the apex of a triangle and controls the other points of the triangle. Relative changes of these two points are similar to the foci of an ellipse, and the position of the Sun and planets revolve around these two centers in an elliptical path, which is experimentally observed much later than the philosophical model. The absolute static is described as Mahavishnu (or Sadashiva) who holds both Vishnu (static) and his consort Laxmi (dynamic) (or Shiva and Shakti). They describe three planes of Atman; one is constantly changing or transforming, one is not changing (static), and the third is the superposition of both changing and static phases. He is playing with living and death. This model is similar to the chicken and egg problem; the egg has two yolks, i.e., two centers that do not touch each other due to rotation. The egg becomes a chicken, and the chicken reappears as an egg. Although these centers are moving, the movement is necessary in order to focus on a unique point as two eyes do. The focal point is also unstable, which is dissociated into many points soon after. This is a purely holistic model of the universe developed a few thousand years before the era of modern science began. Mind is the central part of the model, which has three characteristics: quiescence, attraction to the center, and wandering types like protons, neutrons, and electrons in the surroundings. This concept explains the stability of a system in general. The dualistic nature of the creator, Brahman is described as material type (Apara) and conscious type (Para). The creator is described as a half-man and a half-woman which is a combination of the Sun and the Moon. Sun is the soul or Atman working with the Moon as the mind. The Earth represents our body, which contains life.

The Sun emits different kinds of radiation in all directions; however, a single model having a pole structure of the universe can be developed. To explain the emitted field lines in all directions four directions like the four heads of the creator Brahma can be constructed. In a two-dimensional plane, it looks like the lines are aligned in left and right directions with a line that is straight in the upward or downward direction. However, two coupled magnetic centers make the picture of the universe more complete. Because the field lines emitted from one magnet enter into those of the other, a closed world is made even by inserting more magnets. Ultimately, it is the Sun that produces magnetic fields, which are *Maya* and different from electric fields or light or heat. Such a magnetic field can be imagined with the main deity (God, Vishnu) with two assistants in addition to his consort, or Krishna-Radha may be compared with the north and south poles of a magnet. The same model can also be applied to the electrical fields. Now we try to develop a connection between the old terminology used in the scriptures and possible explanations in modern science.

Vortex

A vortex can be described as a combination of electric and magnetic fields that develops a spin-orbit-like coupling. Since a vortex is created by space–time distortion, it is a playground for studying disorder. Thus a holographic (multiverse) can be created not only

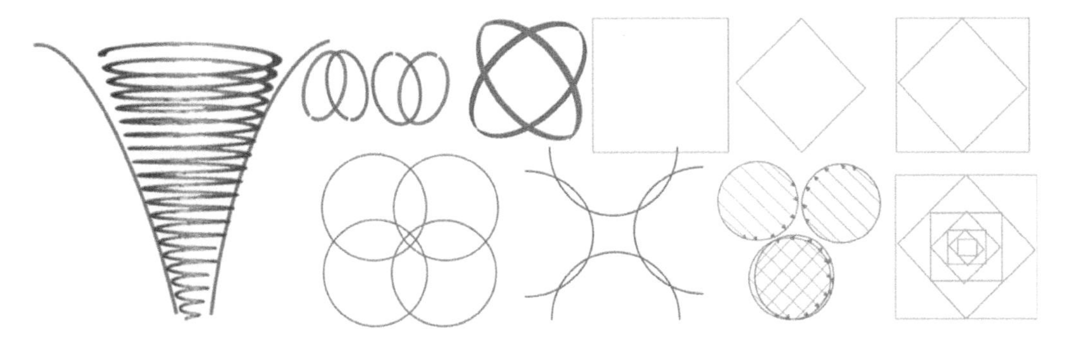

FIGURE 0.9
Curvature of space created by a massive black hole. A conical shape or a vortex structure leading to an infinitely deep potential can be formed. The structure can be described as a black hole in deep space, which is formed by the multiple folding of space. The space is created from a spiral. It is filled with two or four objects oriented in opposite directions e.g., two ellipses, four overlapping circles, and two sets of squares. The squares are concentric and are aligned at 45 degrees.

TABLE 0.1
The physical significance of the philosophical terms

Philosophy	Significance	Philosophy	Significance
Indra	King of gods	*Purusha*	Absolute static, localized, Sun
Atman	Soul	*Vishnu*	The preserver
Agni	Fire, acceleration	*Shiva*	The destroyer
Vayu	Flow, current	*Prakriti*	Dynamics, current, Moon
Brahma	The Creator	*Gunas*	Variable characteristics, colors
(Maha) Vishnu	The (great) Preserver	*Sat (Satwa)*	Fixed, bound state, nucleus
(Sada) Shiva	The (ultimate) Destroyer	*Chit (Rajas)*	Mind, waves, rotating, electron
Shakti	Primordial energy	*Ananda*	Joy, attraction, condensed state
Indriya	Senses (king of gods)	*Ahamkara*	Ego, Superposition
Brahmanda	Cosmic egg	*Vikriti*	Disorder, nonequilibrium
Param(atman)	Great-soul	*Maya*	Illusion, phase
Param(anu)	Atom, molecule	*Mukti*	Liberation, light, radiation
Linga-Yoni	Vortex–antivortex	*Susupti*	Deep sleep with dream
Para-apara	Wave–particle	*Tamas*	Inertia, darkness, repulsion
Devi	Goddess	*Kali*	Shakti-Destroyer
Chakra	Cycle,phase	*Maya*	Illusion
Mahishasura	Demon in dual form	*Maa*	The Mother
Sunya	Void, black hole	*Darshan*	Vision, philosophy
Mahat/Buddhi	intelligence	*Chakra*	Cycle, phase

by theoretical means but also by experimental simulations. This can connect the classical and quantum worlds. A vortex can range from a microscopic scale starting from a point to a macroscopic scale that is divergent and expands on the astronomical scale. Two vortices can merge and interconnect through tunneling, forming a wormhole. See Table 0.1.

1

Creation? A Cosmic Model—Birth of Matter and Mind: Two Cycles

Each step in wave-like life faces a resistance, and by overcoming it, we taste the beauty of life. To initiate, maintain, and stop the motion, a wave has to overcome static or dynamic inertia. Are there techniques for how to tackle resistance? Creation involves the transformation of energy to mass or vice versa. It can be compared to the conversion of liquid to vapor. The development of knowledge is based on our efforts as practiced in our childhood. The maintenance of our energy is based on the assimilation of energy to mass or masses like a collection of water particles to form a cloud. We overcome any force that tries to stabilize this system. This is like the absorption of energy by the creator. Finally, when a critical mass is achieved, it will dissolve into small parts like raindrops. The primordial energy represented as Shakti (Mother Goddess, Kali) gives birth to many other forms from her body. It resembles an egg producing many eggs, a cycle producing many cycles.

The creation starts from an egg (model), which was in the form of a seed of life at the initial stage. It was not manifested and hidden. Once it becomes visible, it can expand, it can rotate, it can change its color, and it become massives, described as Sattwa (or manifestation of space), Rajas (action or rotation), and Tamas (mass). However, all these changes are assisted by heat or temperature variation. When the egg is formed, it wishes to expand and then explodes. Matter consisting of an egg is regarded as a combination of Sattwa/Rajas/Tamas. For example, the hard shell as the cover: Tamas (the yellow or red part as a life surrounded by the white/transparent liquid as Rajas and Sattwa). The hard shell reflects the heat to keep the egg warm and gives birth to life, which is liberation. The white part keeps the balance between the red center and the shell. The balance between the attractive and the repulsive forces keeps the egg alive and protected. Surely there is a feedback of heat that circulates in the closed space. But the most important is the heat that is Atman or life (or the Brahman). It is everywhere, even outside the egg, but the control of heat (temperature) saves that life or gives birth to temperature: the condensation of heat-into-matter or water-to-rain cycle.

The egg model is a very general one, and it applies to other materials and even to our minds. When you make ice from water, the surface of the ice must be isolated; otherwise it will melt into water. We need a shell or protection that creates a temperature or pressure difference, i.e., an equilibrium. But everything is controlled by heat or the Brahman. As a result, the attractive and repulsive forces cannot be seen but the heat difference can be. So condensation gives duality, a mixed phase of solid and liquid or three phases (solid, liquid, and gas) or five phases (solid, liquid, gas, plasma, and vacuum). The world is formed by a combination of deficiency (vacancy) and sufficiency, which are the reflections of each

DOI: 10.1201/9781003304814-2

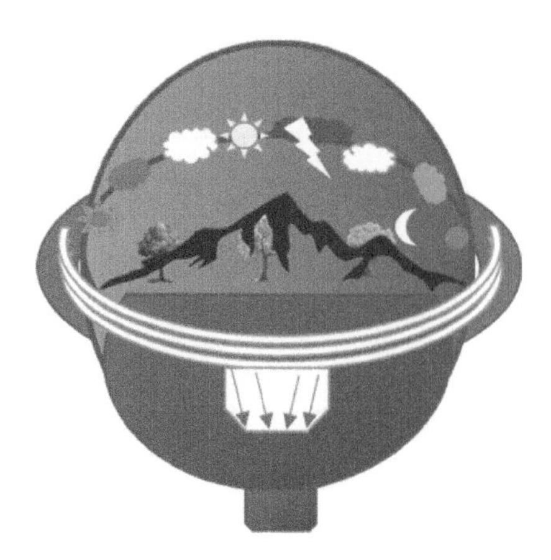

FIGURE 1.1
An ancient model of our planet shows a spherical egg-like shape (atmosphere) surrounded by water and floating on the water bed (ocean). The rivers are flowing around the land, which looks like a bed of water.

other, a superposition of both real and imaginary objects like snakes and rope. The mind can be free like still water, and it can concentrate on a point like focusing on light or the condensation of water into solid form through the transmission of heat at different parts of the body and mind.

Our knowledge reflects a force that is a true representation of Bramha. It can be replaced by heat or shapeless fire. This quantity breaks the dualism into a pure state from the superposition of states. The vacuum of the sky or space is the most important part of the formation of a solid that consists of space between atoms. This space offers the interacting potential, which is heat energy. This is Brahman, who controls everything as a form of space or heat. By changing its potential, a piece of a solid object is converted into liquid, gas, or plasma. Potential stays between and outside any object since it is the only universal object that has no shape. Heat is inside and outside the pot or a potential well.

Although there could be some repulsive forces in nature, it is understood that for equilibrium, there should be a mechanism for creating an attractive force that can absorb everything. It is like a wish of Shakti to consume everything and to try to find what is lacking. Finding this contrast, i.e., excess and deficiency standing side by side, is the origin of 'creation'. The attraction attains a maximum when it reacts with its counterpart or image. This is the state of ultimate knowledge when an equilibrium object creates a resonance, which is regarded as the first sound.

We experience attractions in different parts of life that always play a key role. So how do we define an ultimate attractive force when it is related to god, a very strong force which draws our attention? There must be a purpose for creation, i.e., completeness. The created object should be beautiful (no incompleteness would exist); i.e., it should maintain equilibrium. There is ongoing debate about whether someone created the Universe or the Universe was created by itself. In both cases, the purpose or cause is the same: reaching an equilibrium or achieving Beauty. This is obtained by nicely placing objects of high contrast. The purpose is to create a 'beautifully decorated disorder'. It is like creating light in the dark. A flame looks nice and complete when surrounded by darkness. Look at the moonlight at night, the stars are unappealing or invisible but not so on a night when the Moon is fully covered by darkness. One type of contrast is not enough. We need three

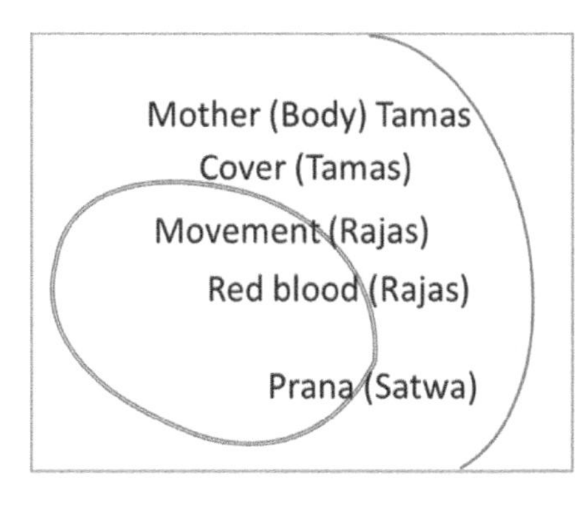

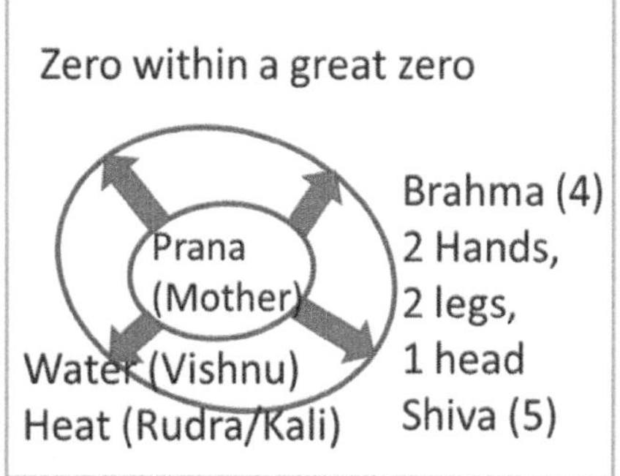

FIGURE 1.2
We are born from the ellipsoid shape of our mother's womb. Our small body floats in the water before our birth. The outer layer of the womb filled with blood and water, which represent Tamas, Rajas, and Satwa, respectively, helps the life to grow. The body carrying life remains in a dormant state, which can also dream. Slowly our body is developed. Two hands, two legs, and one head are unfolded from an egg. This unborn child carrying life represents three prime gods: Vishnu floating on the water bed, Brahma having four heads, and Shiva having five heads.

items to coexist. Like an equation: one is produced from the interaction of two, which also maintains contrast. This is a simple form of creative force or tension that looks appealing or effective when some kind of wish or acceleration fights inertia (resistance to force). The energy is manifested beautifully when mass (or inertia) is competing with time which spreads over a space. At first, one odd number is created, followed by another odd number having an opposite parity to form neutrality. Similarly, particles that rotate a complete cycle or a half cycle that can have opposite characters are created. Finally, particles with fractional properties and a mixture (a state of disorder) are formed.

Developing Vision for a Common Observer: Creating the Space and Cycles (Matter and Wave)

We extend the concept of vision that was developed in the Preamble. By this time, I know how to create and manage fire and how to secure my food reservoirs using various devices. I know some concepts of force and the shape of the sky and curved space. However, I do not know how to combine natural and physical processes that can be modeled based on a most common device, i.e., the wheel. One day I think about how to develop ideas and intelligence. My query was how do we learn from nature through interaction, and how does a large idea develop from a seed of knowledge? I looked at the first discovery of man, i.e., fire, and then some weapons made of stone and wood or bone used to kill animals. After collecting 'water' or grains of food or fruits, a kind of pot was found to be essential. So space is created from the flat surface or mud in the form of a closed curved space, a cavity, or a void. To create a circular or spherical space, a periodic motion is necessary, which is provided by a wheel. The wheel has a fixed axis of rotation, and the rim of the wheel circles around the axis or interacts with the axle. The outer part of the wheel obeys

FIGURE 1.3
Resonant sound is created from a cavity by a rod.

the axis, which is virtually inactive and holds the whole process together. But you need an external force to initiate the rotational motion and to continue the process. Along the axis of the wheel, a linear motion is produced, which can also be considered a helical motion or the evolution of motion. So from this elementary oscillation, you can create an extended rotational system like a gear system connected to a motor.

Potters use this technique to create the desired shape of a vessel from a lump of clay that is inactive like a dead body. From a large piece of hard/solid stone fine grains of soil mixed with water to create a soft, flexible material that can be shaped and hardened. The mind controls the shape of the potter who has the vision for the desired shape and who applies pressure in the direction normal to the axis of rotation at different distances. This is torque or moment, which controls the size of the emergent body. The moment is the mind, which controls the whole process. This develops the idea of Chakra (wheel) in our bodies. From the stomach, heat is produced and goes upward in our body and supplies energy to the brain. Energy becomes maximum through its passage via wheels of different sizes and shapes, like a gear system. It is like expansion and contraction. From the invisible shape of mud, a shape is formed. From the silence, the sounds of different frequencies are formed through expansion and contraction.

The world has a round/sphere/oval shape as we would like to see curves instead of lines. On our flat body, we have a round face, which is beautiful, instead of a triangle or square. Hot vapor inside a flat body inflates it into an oval or round body. Our human body is created in this way. A slow change of curves creates a rounded figure, which is a symbol of completeness, stability, or satisfaction. Looking at one or a couple of round objects gives us enormous satisfaction, it excites us, inspires us, and also attracts us to meet the round objects. If there is some incompleteness, a gap lets us fill it up and makes it more complete. God played with round objects from a dispersed gap to create a round object, which gives an order defined by the area or radius of the object that passes through a dot and becomes large again. Focus can lead to an order (from disorder) that needs a very strong force to unite all the scattered particles. If we start from three lines oriented in different directions, they can form at least a triangle. The area or length of the triangle is a measure of order. The centroid of a triangle gives an order similar to the center of a circle. Ultimately the triangle goes to a point and rises again. Focus is most important in giving duality around the forces. Duality needs a focal point.

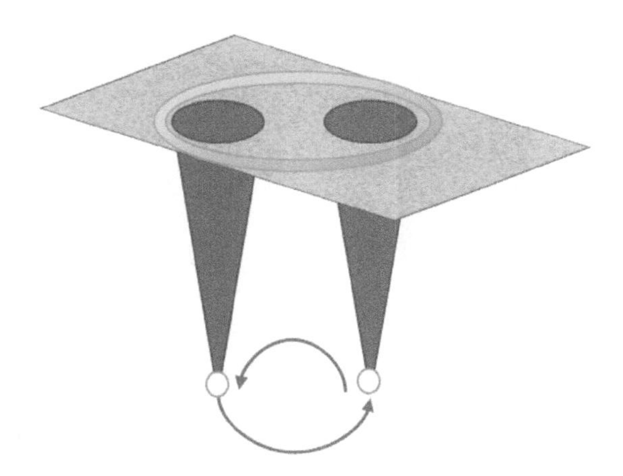

FIGURE 1.4
The cycles are created by two massive stars. Two deep potentials are interacting that produce gravitational waves. Two funnel-like vortices are exchanging their positions continuously like flipping two fingers. This is an example of dualism.

When we are alone and watching the night sky filled with stars, we develop a feeling that the stars are like eyes against the night that are constantly watching us. Thousands of stars mean thousands of eyes that are pointed at you. They form some patterns, and they move throughout the year. But you can trust them because they do not go away but reappear every year without fail. During the day, the Sun is also watching you. At midday, you are taking a rest under a tree with a thousand leaves, and sunlight appears to you as a thousand small light rays. It is difficult to escape under the tree from the Sun that is watching you with his thousand eyes. Thousands of arms like the Sun rays are hitting you. One Sun god with thousands of heads, eyes, or hands is Brahman. We need a unified field that contains all. From one flame of fire, you can see that many flames are created. One head or tongue of fire is split into thousand, like many horns. One Brahman can have many Atmans. Even water from one source (ocean) forms thousands of raindrops on your head. A unified field is different from a discrete one. Initially, the Universe is described as many stars in a circle of light that as a whole is in god: Brahman, like the king of gods Indra. But we wanted to have one god with the minimum splitting of forms. So Indra (of many heads) is replaced by Vishnu together with Bramha (four heads) and Shiva (five heads or elements), who can provide a universal field of consciousness that is spread everywhere. Discrete stars become a single body under one energy, Vishnu.

Focus is like a singular point. Beyond focus, we cannot see anything but uncertainty. Focus is the center of man. All energy is concentrated. Let's consider that an object (man) is moving at the speed of light. My eye cannot focus on an object moving at that speed, and we cannot see the photons. Photons do not have any mass, so no energy is spent for light to move in straight lines. But if we add mass to a photon to keep it away from our focus or viewing distance, then it would cost energy. A space is defined as c^2 (a two-dimensional space, like a square) that will move at a speed (c) that cannot be seen. To move the space (c^2) or the mass (m) out of our vision is equivalent to energy E. Since energy is concentrated at the focus, this is also described as a collapse of space to a point, i.e., $c^2 \rightarrow 0$. Since c is large, the space is curved. Space is a vacuum, which is the inverse of matter inertia. In the representation of the Universe, Vishnu changes from inertia to space and creates Bramha, the energy or light created. Mass changes man to space and makes light (sound). Moving

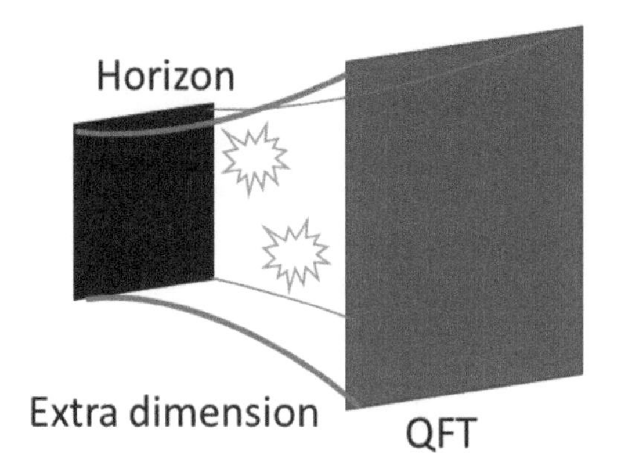

FIGURE 1.5
The Universe created from a black hole is expanding continuously. From two dimensions, a three-dimensional space emerges. This shows the duality of gauge and gravity. Duality creates an extra dimension. The increase of dimension looks like a flat top pyramid with the smaller base representing a lower dimension and the larger base representing a higher dimension.

the cloud (mass) at the speed of light, like the Sun's (Vishnu's) rise, creates energy (light) or Bramha. Breaking the dream of Vishnu gives vision 'focus', which is Bramha. This process is described in detail in Chapter 3. Straightness like that of an arrow has a target. If the person rotates the stick in all directions, then it is a chakra.

The Sun and the Revised Egg Model

Even today, as we observe the motion of the Sun, the brightest object, and our loving Moon (and the stars) moving above our head every day and night, we can see they are in an imaginary curved space like a dome. It is easy to believe that some kind of force is trying to pull them toward a center, i.e., why they are in a curved space around the observed (fourth element/Earth/matter/reference), yet they are not falling or merging to the observer. That means someone is holding them from the top (heaven/god). So there is another center in the sky. It could be the Sun that we are afraid of due to its bright light and heat. It has also a round shape. This attractive force can be described by a bow-and-arrow model where the bow provides the attraction and speed to the arrow, which represents the direction. Creating attraction on the arrow bends the bow, and the released arrow moves in a specific direction. So the curvature of space has a relationship with rotation that gives stability to the system. This idea gives birth to 'wheels', which have a fixed center and which attract to the center. By rotating the object on the rim of the wheels, you get speed: the conversion of linear to rotational motion. This is the bow of motion. But these were devices (the bow and arrow, wheels) that are human-made. What about big or very small wheels that do not have a regular shape or size? How do we explain that nature consists of fire, water, air, and matter since space is not empty but filled with these four elements? All these elements maintain a cycle. They represent plasma, gas, liquid, and solid. Our body is made of solid, liquid, and gas (air) for the exchange of heat and (sky). Mind is infinite, like the sky. The

FIGURE 1.6
Meditation as a common practice was found all over the world in ancient times. A sun-like god is meditating and gaining inner energy by overcoming disturbances and rejecting noise. The head is accompanied by two structures.

Sun and nature (Shakti) are represented by several prime deities, gods, and goddesses in Indian culture, as discussed next.

Creating a Creator and the Prime Gods

I am the god with four hands or two hands plus two wings sitting or an animal as the carrier. One is active while another is passive. Ideally Purusha is completely inactive, and Prakriti'is very active (not static). However, the Purusha is constantly watching Prakriti at his focal point of vision. This is like a bird that is constantly looking at the egg that she laid. Initially, the Krishna–Radha model in Hinduism was developed based on the interaction between the Sun and Moon. Similarly, by combining Shiva and Kali, a model has been developed where Shiva is completely inactive but looking at his consort Shakti (goddess Kali). Kali contains all three characteristics (the Gunas).

I, the creator Bramha, can see my creatures with my four heads. In the daytime, I watch Earth and set my feet on these places at dawn, noon, and sunset. I am Vishnu. This is my stable structure (at equilibrium). Ultimately, I am Shiva (Rudra) who is silent but responsible for destruction, i.e., the beginning and end of a cycle of defining time and the ultimate time. The world is 'round' as we see it every day: Why does 'the sky' look like a dome on our head? We search for infinity. Our eyes try to focus on space. If successful, then they can see something at the focus. But if the object is too far in the sky, then we cannot see it because it is out of our range, the maximum distance that we can see. If we join the focal points, it forms a dome. The range we see is the 'space', like a bowl. Our eyes are not the best tools for measuring distances (at least large distances). We cannot distinguish between stars (or planets) by distance easily. We think all stars are on a spherical surface that covers our planet or vision, i.e., Maya. Reflection of this imaginary plane gives different colors as well. Focus means can create matter by concentrating fields. Focus is the most important thing because it creates this virtual space. If an object is close by or within our visible range, then a straight line looks 'straight' absolutely. If the length increases, then it can go out of focus at the ends. So if you look at both ends of a long rod, it starts bending. If the line is infinitely large, then it is seen as a curve. This is still a classic picture without considering the speed of light.

The Sun represents Atman, which remains static. Sun can be seen from the Earth as red, white, and red, but that is unreal; only the cycle of the Earth is real. It is like sleeping over the night, getting out of bed at night, and going to bed at the end of the day. The superposition happens between bright and not so bright Sun due to its relaxation. The Sun appears to us like a ball of plasma. Sattwa/Rajas/Tamas is equivalent to being awake, asleep, and Susupti (meditative). About the combination of colors, ancient philosophers described fire as connected to the heat source, which gives different frequencies by controlling the heat to different temperatures. Fire has been classified into three different temperatures that give five different frequencies. Fire is the prime god in ancient cultures and is responsible for creation.

Agni from the sky observes us, and we observe the creation of Agni. Through observation (our eyes, ear, and all senses), we merge into observables becoming one body, I and He. If we wish to see him while our eyes are closed, He comes to our mind in different forms mostly to solve the problems that we are dealing with. Sometimes, several (three) light beams converge at a point to form a bright object that blasts (in a big bang) into many small particles (or bright light particles). This bright egg-shaped material is described in scriptures as 'Bramhanda'. Based on this idea, modern cosmology is trying to present a possible model of the Universe and the formation of the stars. Afterward, the big bang particles move in a straight line or follow a curved linear path. Now we enter into the realm of logic: Is our world flat or curved, and, if curved, then what kind of curvature? The geometry of space has started. We understand that the world is not a flat space. The prime gods are described next.

The goddess Devi was the first creation. In the ancient texts, the supreme creator was described as a man-like or a woman-like character. The prime mother goddess Aditi gave

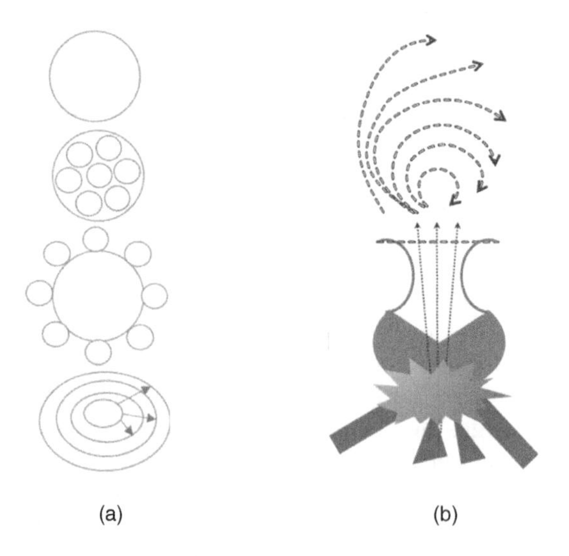

FIGURE 1.7 (a)
From a large circular body, many small bubbles are created. They form multiorbit bodies such as atoms. A large circle encloses several small circles or is surrounded by smaller circles, and also a set of concentric circles.

FIGURE 1.7 (b)
The water cycle represents a heat cycle created from Agni, or fire. The cycle looks like an ellipsoid. The circular shape of the vapor produced from a pot of water placed on a fire.

birth to everything including the Sun. She is described as 'infinite'. She created one god Daksha, who also created Her in return. The creation is therefore described as a feedback (loop) process. The first creation, the female goddess Devi, arises from the body of a man-like creator. He does not have any birth or death and has not been created (or created himself). However, He lives in a dormant state and does not influence Her activity. She plays with duality and gives birth to three sons: Bramha, Vishnu, and Shiva, three prime gods. The female goddess can be formed as a synthesis of three different characters or three male creators. I was born as an EVEN number (a duality), which consists of two ODD numbers. Both parts of mine were created simultaneously as a pair (a dual). Interactions of these two parts produce an ODD number, such as one or three.

So the space is created but not the energy or time (Shakti). Shakti gives birth to sound (a wave). Sound consists of our words, Buk (described in Chapter 3). She is Bramha as described in *Rigveda*, 10th Mandal, 125 Sukta as the goddess Chandi (in Sanskrit: 'The Fierce'). The universal mother consists of three prime gods, namely Vishnu, Bramha, and Shiva. From the life of energy (heat), sound energy (Buk) is created. It has four forms; three of them are not manifested (nonobservable) and are manifested as the sound used by humans as dialogue (conversation). Agni or life energy is the husband of Buk, sound who is also Agni Devi, Ila Devi, Saraswati Devi, and Bharati Devi. Three fires and one. So one can imagine that all the stars were together in a small body that had the power of all stars combined, i.e., it was extremely bright. After the bang, Bramha became many.

We have understood that God at one time created heat, light, water, space, and energy, which are static and hold the world in equilibrium. But this is the time to create Time and Frequency, which have different periods and which can be done through sound, particularly with Air. This is also Fire, a representation of the Sun that creates excitations within itself through interactions (nuclear) that produces light, heat, and fields (magnetic). This mechanism can be applied to produce fire on Earth through the friction of wood, which creates a high frequency of particles and then a spectrum of frequencies of light. In our body, air intake with the inner parts produces heat. Without breathing or the exchange of energy, we cannot survive. In our minds, also, when we concentrate, all energies are united in a very controlled manner, and they are centered in the brain and create fire or a very bright object, which can last for some time and then decays. This is God observing the fire in the mind. It seems to be that the interactions of bringing all energy to a particular point in the brain creates a charge or polarization through alignment. Without the alignment or polarization of matter, a condensed state (a concentration at some place) cannot be produced. This gives us consciousness.

So from many thousands of stars of energy particles, one dense ball of fire is produced, and it dissociates back into many bright particles. A cycle is completed. The Buk goddess is of four types, which represent the four actions of creating frequencies. Three of them are light (no visible or audible frequencies) and one of them is manifested as audible sound. The unified one is Bharati Devi, which gives us pure thought (consciousness) overcoming mental instability and contradictory ideas. Bharati Devi encompasses the whole Universe and is Para Buk. Durga Devi, or Bharati Devi, lives inside the Sun and creates fire through nuclear interactions. The energy of the sky is converted into consciousness energy in the sky ('Breaking up'). Mahalaxmi Devi (or Ila Devi) uses sunlight to excite water, which interacts in the sky (clouds) and is dissociated by lightning (in a charge), giving rain. 'Breaking up'. Conversion of light to heat produces consciousness on the planet.

Mahasaraswati Devi is the conversion of solar energy into human brain energy through the interaction between the Moon and Earth. Bharati Devi lives in heaven and signifies interactions between the Sun and the Earth's surface, which creates our voice and related

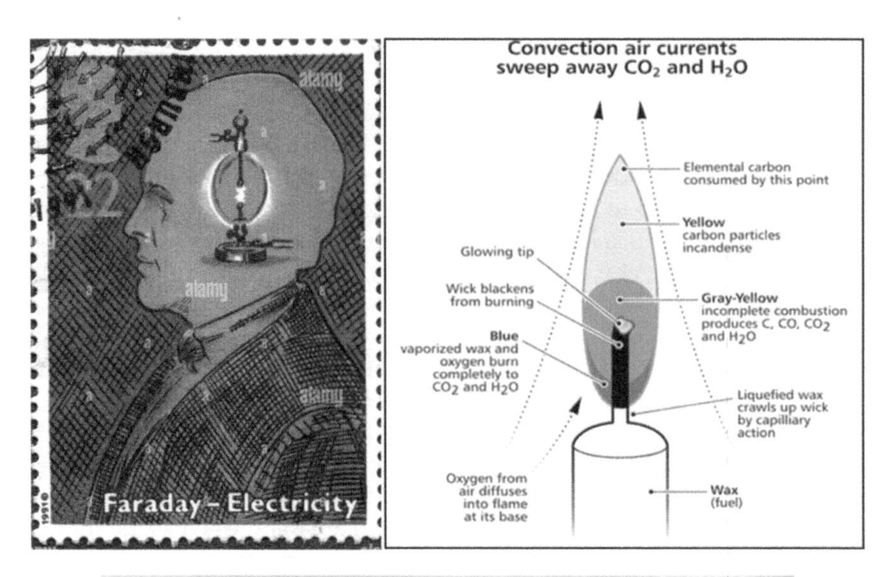

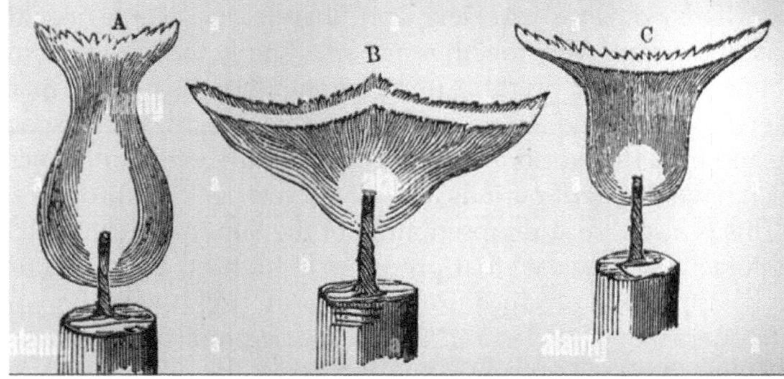

FIGURE 1.8 (a)
"There is not a law under which any part of the universe is governed which does not come into play and is touched upon in the chemistry of a candle," writes Michael Faraday. By radiation and conduction, the flame's heat melts the wax; capillary action drags the wax up the wick; the wax then vaporizes; chemical reactions create the flame; heated solid carbon particles glow; and convection currents remove the combustion byproducts.

FIGURE 1.8 (b)
Faraday's demonstration of the effect of magnetism on a candle flame, shoving the result of placing magnet in different positions: A. Candle across magnetic axis; C. Flame raised above magnetic axis; B. Flames above magnetic axis. Faraday, 'Analysis of Flame.

sound. Gravitation is generally meant as the attraction between two objects. The attraction between Earth's surface and the Moon is Saraswati Devi. It is believed that our minds are connected to the attraction of the Moon, which gives us different ideas (frequencies) because the variation in attraction between the Earth and the Moon can create diversity in our minds. This produces the sound from our mouths in the form of language or songs and melody or music. Ila Devi or Mahalaxmi signifies interactions between charges in the cloud or sky with the surface of the Earth. It is manifested as the sound of nature such as thunderstorms.

Aditi is pure consciousness. This gives AUM: A (creation: Saraswati), U (stability: Ila, Laxmi), and M (decay: Kali). This female goddess can search for her male consorts Bramha, Vishnu, and Shiva.

Summary

This chapter addresses the goddess of Shakti or Kali the destroyer. In the previous chapter, the Sun (or Brahma) is described as the prime god, and Agni is created from the Sun. In this chapter, we present the concept of time and phases, which can be connected to the creation model. How did the concept of deities develop from nature including heat-light, water, and air? Agni is the main god in this chapter and is manifested in different forms (or colors). Agni, or fire, gives us a vision that is created from light and the interaction of matter. We discuss the creation of gods or the creators. Most importantly, we claim the importance of gods in the female form in creation. Goddesses can create the prime gods and fight against demons.

In the following chapters, we see one goddess (Chandi) who fights against the demons who appear in pairs. She kills the demons finally and regains equilibrium in the world (nature). The battle between the goddess and the demons will be elaborated on in Chapters 7 and 8 in the context of the vortex and quantum fluids.

References

Brahma: From Prajapati who created land to hold the cosmic water. Visha Karma asked for space for living on Earth, and then Prajapati created sky or air. Both are Sun, actually, which is also Indra and named Brahma. His body has two parts like a fly (duality). With Trastr, they form three main gods of creation. Brahma is created as a lotus, which is a symbol of the Sun that originates from the *Rigveda* (10th Mandal) where Visha Karma is described. It is cosmic water where all the gods live and meet. It is like the navel pit of a Purush, where Brahmanda exists. Visha Karma lived in the water and created life. So the story of Vishnu started from Visha Karma and from all the gods who lived in the navel pit and became Vishnu in Purans (the Padma or Brahmanda Puran).

Vishnu: Gravitational force stabilizing the world. We take this story of Vishnu directly from *Markendya Puran, Sri Sri Chandi,* where Vishnu is described as 'Narayan' or god of water who sleeps on the bed of water, i.e., in the Universe or space as the Sun. Brahma had a competition for superiority. Vishnu entered the body of Brahma and saw three worlds. However, Brahma entered Vishnu's body, saw an infinite world, and escaped through Vishnu's navel since all the other exits were closed. It was said that Brahma was born in a golden egg floating in the great ocean. He created a water-like space and then the Sun. He divided the egg into sky and Earth. However, he is considered a Narayan or at least part of Vishnu, i.e., the Sun.

Shiva: Rudra is created from the forehead of Brahma. He signifies life. Like Vishnu, he was also produced for Visha Karma's body on a lotus, which can be the Sun or Earth. He is Agni. He originally had five heads but now has four heads, signifying four basic elements or two pairs of opposite elements, i.e., +/− charge, the north–south pole of the Earth and Sun. One neutral point signifies the sky or the intellect. However, that head was recorded by Shiva and put on his body, so Shiva became five-headed. Vishnu has three legs and we shall explain this later. He represents the Sun and fire.

Kali: She breaks up space (s) into fragments with time (t), which is $\mathrm{d}s/\mathrm{d}t$ = velocity; however, if space is considered as something moving at a constant speed, then $\mathrm{d}v/\mathrm{d}t$ = acceleration, which gives force or reaction. Space means that something is gliding at a constant speed. Kali can overcome time. Kali is timeless, infinite, singular, unmanifested, and formless. She is the main cause that is hidden inside time, including the past, present, and future. Kali yantra consists of eight energies (different from Sankhya): Bhaihmani, Narayani, Maheswari, Chamunda, Kaumari,

Aparajita, Barahi, and Narasimha. It is also surrounded by eight elements: Earth, water, fire, air, sky, Moon, Sun, and the priest. It is surrounded by eight elemental objects: Earth, water, fire, sky, mind, intellect, ego, and infinity. She is supported by eight Vedic gods: Rudra, Basu, Aditya, Indra, Vishwadev, Agni, and the Aswani couple, i.e., Mitra and Varun.

***Markandeya Puran: Devi Saptasati: Sri Sri Chandi*:** This is one of the greatest books that I have studied in my life. This book speaks about overcoming bondage or knots of life or the complexity of space. In the first section, it is overcoming the knot related to Brahma showing Mother goddess as a very strong introvert force that produces Earth from cosmic dust or plasma-like water. The main obstacle was two demons Madhu and Kaitava, who stopped the energy to fall onto Earth from the Sun. The second part is the Vishnu knot, where the gooddess Devi is created from the Sun by concentrating his power to a point to create the Moon. YMoon was not stable, and the phase was changing. It was described by the demon Mahishasur, who changed from one animal to another. The third was associated with Rudra for the creation of the Sun itself. I am still understanding this complicated part.

The first part of the book describing the Maya of Vishnu will be discussed in various chapters. Vishnu lies on a water bed and is dreaming but could not create matter or Brahma due to the noise of the demons Madhu–Kaitava, who are also parts of Vishnu. Vishnu becomes angry and releases Kali from his body to fight the demons. Maya, or the dream of Vishnu, is overcome by Mahamaya (or Shakti) when Vishnu rises. Vishnu kills the demons with a weapon called the Sudarshan Chakra (a great wheel) and created many parts like stars and planets from the state. This is said to be the rebirth of Brahma. Vishnu was dreaming and did not want to be disturbed before creating Brahma. This part will be discussed thoroughly in Chapter 3.

Durga, who is white, created Kali, who was black and who wishes to have a world of harmony. She kills all other dualities that are represented by demons in pairs like Madhu-Kaitava (attractive and repulsive forces), Mahishasur (the changing phase of different animals), Chanda-Munda (spheres made of wish and mind that are unstable), and Shumbha-Nishumbha (holes made of good and bad things from Kumbha, which means vortex or antivortex). As deadheads, these demons are associated with Maa as a garland where the heads can be considered a planetary system. One main orbit of the string is decorated with small orbits or many garlands. This is the only model of the Universe we follow. This will be discussed thoroughly in Chapter 6.

2

Background/History—Understanding Creation: Gods and Goddesses

I begin this chapter with a story describing the battle between the Devas and the Demons. Indra is the king of gods in this chapter. Just as the goddess (Chandi) fights against the demons, Indra fights against one demon who is killed ultimately. The story follows.

Battle between Devaraj Indra and the Demon Vrittasur (Battle of Two Cycles)

As per the ancient literature and as contained in *Shrimad Bhagawat* as well as other religious books, there was once a Prajapati named Tvashta. Prajapati Tvashta had a devout and pious son named Vishwarupa. Vishwarupa is blessed with three hands. He was a sage having immense spiritual strengths, which evoked a sense of insecurity and fear in the mind of Lord Indra, the king of Indralok, or paradise. Once, in a fit of rage, Lord Indra killed the good sage. When Prajapati Tvashta came to know about the incident, he became furious and performed a ritual in front of a sacred fire (Yagna) to avenge the death of his dear son. From holy fire, another son of Tvashta was born. This son was named Vrittasura, whose sole aim in life was to avenge his brother's death by destroying Indra. The word 'Vrittasura' means the 'enveloper'. The slaying of Vrittasura is considered the greatest achievement of king Indra.

Vrittasura then meditated and undertook a penance, as a result of which he was granted a supreme boon. As per this boon, no weapon was known till then that could kill Vrittasura, and Vrittasura would not die even of anything that was either wet or dry or any weapon made of wood or metal. Vrittasura had kept the waters of the world captive; therefore, Indra planned a battle against Vrittasura. Before going to war, he drank a large volume of a sacred wire (Soma) at Tvashtri's house, who is an artisan god. Tvashtri designed the thunderbolt for Indra. Then Indra requested Vishnu to create a space for the war. Vishnu did this by taking three strides and hence called 'Trivikrama'. On being granted this boon, Vrittasura waged a battle against Indra and his forces and finally managed to give Indra a crushing defeat, as a result of which Indra had to flee from the battlefield leaving behind his elephant Airavat. Vrittasura broke the two jaws of Indra, but then Indra threw him down, crushing the 99 fortresses of Vritra. All the gods under Indra's leadership attacked

DOI: 10.1201/9781003304814-3

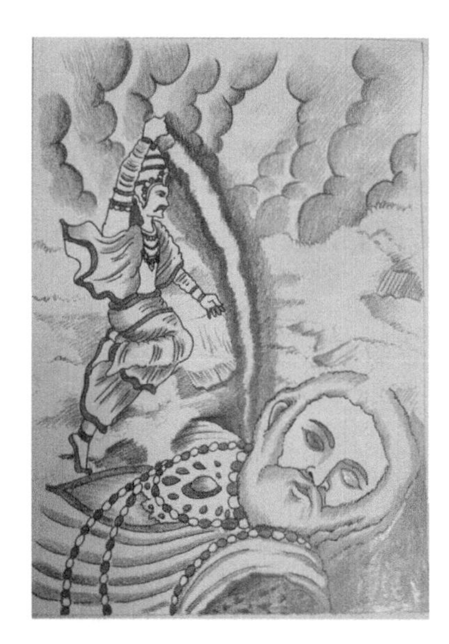

FIGURE 2.1
The king of gods, Indra, destroys a demon-like character (cloud) with his weapon (lightning) to produce rain.

Vrittasura with their weapons, but Vrittasura swallowed all their weapons. The fearful gods then ran away and went to the refuge of Vishnu, where they eulogized him. Vishnu, being praised with their eulogies, advised Indra to approach the sage Dadhichi because his body and particularly his bones are stronger than any weapon. Therefore, only a weapon made from his bones was useful in killing Vrittasura.

Accepting the requests of gods, sage Dadhichi gave up his body for a great cause and his soul left this world. With the help of Vishwakarma, the gods prepared vajra (a thunderbolt) from the bones of sage Dadhichi. Armed with this vajra, Indra and other gods attacked Vrittasura. This war took place at the end of the Satya-Yuga (period) and the beginning of the Treta-Yuga. Although one of his arms was severed from his body, Vrittasura angrily approached Indra and struck him on the jaw with an iron mace. He also struck the elephant that carried Indra. Thus Indra dropped the vajra from his hand, but Vrittasura let him take his vajra again and fight. After this, Indra cut the remaining hand of Vrittasura. But the demon did not give up. He assumed a gigantic form and swallowed Indra along with Airavata. With vajra, Indra pierced through Vrittasura's abdomen and came out. He then immediately cut off Vrittasura's head, which was as high as the peak of a mountain. Although vajra revolved around the demon's neck with great speed, separating his head from his body took one complete year—360 days—the time in which the Sun, Moon, and other luminaries complete a northern and southern journey. Then, at the suitable time for Vrittasura to be killed, his head fell to the ground. After Vrittasura's death, a living spark then came forth from his body and returned home, back to the Godhead. He entered the transcendental world to become an associate of Lord Sankarshana (a form of Vishnu).

This story shows the establishment of equilibrium from three elements that accumulate enormous energy and destroy the duality. Duality in the sky creates a charge separation in the cloud that looks like a massive (black) demon. By applying an electric field (lightning), Indra breaks the duality, resulting in raindrops.

The Vedas

An ancient source of pure knowledge of nature can be found in the Rigveda. This vast text often addresses trinity or the knowledge of trinity, which is divided into three (or four) parts: the *Rigveda*, the *Samaveda*, and the *Yajurveda* (and *Atharbaveda*). It deals with the types of Agni that belong to space (the Sun) as a part of a distant object, the sky (lightning), and the fire on Earth. It is interesting to note that three parts of the *Vedas* contain three different types of content e.g., pure Mantras in the *Rigveda*, songs in the *Samaveda*, and something different from the *Rigveda* and the *Samaveda*. In the *Rigveda* we see ten chapters or ten Mandals like ten directions. I find the *Rigveda* looks like the goddess Durga having ten hands. So it represents the Sun and the rays of the Sun. The *Yajurveda* has two parts, Shukla and Krishna, which mean the white and black of nature, respectively. I think that it represents the phases of the Moon. It can also represent the Sun (which is at its brightest) at midday and appears to be white and the night sky when the Sun is not visible. The *Samaveda* represents the wave that is found as air or water on Earth.

So the *Vedas* describe the Sun, Earth, and Moon or the space between the Sun and the Earth. In other words, they represent three different characteristics: static, dynamic, and neutral. Basically, the *Vedas* describe the origin of heat and its application, how the Sun's rays can be converted on Earth.

How does the Sun control the temperature of the Earth?

It rises in the morning, and the heat of a certain place of the rotating Earth increases until midday. As the Sun is setting, the temperature of the place drops. Apparently, the Sun goes to a maximum height and provides a maximum amount of energy during the midday and then falls to a lower level. I describe this process as the absorption of energy and de-excitation. This de-excitation process gives off the emission of energy in the form of a wave. Hence the change of heat causes waves. The beginning of the Sun, the maximum energy of the Sun, and the process correspond to *Rigveda*, *Samaveda*, and *Yajurveda* as the Sun stays in three forms in three parts of the sky. This heat cycle can change the properties of materials used to make weapons. The annealing of steel (or iron) is very useful for making swords and other weapons. Finally in the 10th Mandal of the *Rigveda*, the *Veda* is described as a person with two heads (duality), four horns, three legs, seven hands, etc., which combines all philosophical ideas described in the entire *Veda*. The Sun god is worshipped by the *Rigveda* in the morning and rises from the dark night. The Sun God stays in the *Yajurveda* at midday. In the evening He is worshipped by the *Samaveda*. The fourth part of the *Veda*, the *Atharvaveda*, was written much later; however, it contains some of the important parts (extensions) of the *Veda* such as *Mundaka Upanishads*. In one of the most popular texts, *Sri Sri Chandi* describes the battle between the gods and the demons. Let me describe some other aspects of the *Vedas*.

Veda or *Rigveda* describes God as the ultimate, unique object that can be present in many places and different forms. He controls everything as 'Ishwar', but he can be described as a different form of God. In the famous Gayatri Mantra (3/55), He is described as a luminescent object that exists in three places: Earth, sky, and the intermediate space as fire. This trinity has been described repeatedly in different Mandals of *Rigveda* in terms of three times (past, present, and future), three rhythms (Gayatri or Tristap), three gods (Agni on Earth, Indra or Varuna in the upper sky, and the Sun in the sky (1/1)). Agni is the most important god in *Rigveda*, which also has three main flames of red, yellow, and blue colors. Agni is prepared from friction between two pieces of wood, i.e., dualism. Two forces working in opposite directions produce a fire-like torque, which causes the rotation of a wheel. However, this process produces the three types of Agni used in rituals.

Nothing is stable forever in the Universe. At some stage, it will become unstable toward death and collapses. From there it tries to attain equilibrium. If it collapses in the Universe, then heat is produced. The heat is also produced when there is no matter. This is like Shiva or Rudra. It tries to break up things, and duality is formed. The two opposite things try to interact with each other and attempt to attain equilibrium. If these things are unequal, then they will compete with each other, e.g., light and heavy. This interaction requires space (the battlefield or playground of Vishnu). The interactions between heat, energy, and space create mass or matter. This will be explained further as the basic concept of the theory of relativity (mass–energy equivalence) in Chapter 5. The interaction, called 'moment' or 'torque', starts mechanically and finally reaches equilibrium (describe the god Brahma).

Yajna and *Homa* are the fire rituals described in *Rigveda*. Three types of rituals are described in *Rigveda*, which are necessary to create a machine that produces a cycle, i.e., the water cycle (1/34). Particularly in 1/34, God is asked to appear three times, and the food offered to God is carried on a vehicle that has three rigid wheels (cycles) and carries three stands. The vehicle looks like a triangle or a pyramid. Everything described can be classified into the number 'three'. This is a Triplet Universe containing three elements, three times, etc., like dawn, midday, and sunset. Agni has also been described as a combination of seven flames. So the shapeless Agni has been given form having different elements with specific numbers that make up a model of the Universe (4/58). In 4/58, he is described as a body with two heads, i.e., two types of descriptions of three and seven elements. Seven comes from a duality or two and five elements. It has seven hands like the seven colors of a rainbow. It has three steps like the three elements described previously. It also has four horns or legs, which signifies the four directions. This structure of "Iswar" is repeated in different parts of *Rigveda*; however, an improved description is given in the last Mandal, i.e., the 10th Mandal. We find the number of Gods as multiples of three, i.e., 3, 33, 3,330, or 33,000,000 (thirty-three million), but they are, like Agni, Indra, and the Sun, in three Universes: Earth, air, and the upper sky.

The most important explanation of the trinity is the three types of sound, i.e., low, intermediate, and high frequencies, that create the whole alphabet and all kinds of phonetics or sounds. In the 10th Mandal, particularly in 10/90, a large Purusha is imagined as having thousands of hands/legs and heads. He covers the Earth and beyond the Earth, like the Sun. His body is divided into pieces and creates different objects and creatures, namely Brahmins from His mouth, kings and warriors from His hands, traders from His legs, and the class of workers from His feet. The Moon is created from His mind, the Sun from His eyes, Indra and Agni from His mouth, air from His soul, the sky from His navel, heaven from His head, Earth from His feet, and different directions from His ears. So He contained all the elements within him. It is not said that He is the creator, but the whole Universe was created from His body. He is Brahma who is unique but has transformed many, like a seed giving rise to a plant having many branches, roots, leaves, and flowers. God separates the sky from the Earth (10/81).

In *Rigveda*, it is also said that Vishwakarma, the creator of the Universe created and maintained the Universe and watched the Earth all the time (10/82). This is like the description of the Sun used in all religions, particularly Christianity and Islam. However, in the 10th Mandal of *Rigveda*, the creation is described as a Purusha that is a form of ultimate consciousness. In 10/12, it is said that somehow heat is created, which is conscious. This object is full of desire, which is the cause of creation. Consciousness added to the desire of splitting creates everything. Thermodynamics and mechanics are the cause of creation. Energy and matter interact with each other to create everything. Energy and space (expansion) create matter. This is the trinity.

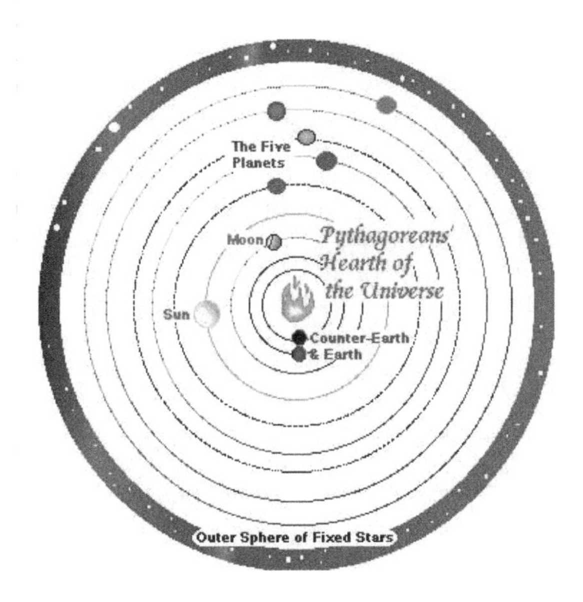

FIGURE 2.2
The Greek philosopher Pythagoras's geocentric model shows many orbits around the center, Earth, and the fire. The fire at the center is surrounded by concentric orbits. Each orbit contains a planet. The outer thick orbit accommodates bright stars.

Heat is created by duality with a maximum difference. Heat at the start increases. Heat is the ultimate potential or kinetic energy. This strong attractive force can bend light. Heat is a product of strong attraction, e.g., friction or rubbing. Heat is an electromagnetic wave (interaction between electric and magnetic fields). Light is created by the heat in one direction. Light is a form of heat used in linear or flat spaces. God (invisible/singularity) is heat/wave/vibration, which can propagate in any direction; hence many heads, eyes, and legs.

Heat-Fire-Light Rudra: Agni

It can destroy and transform everything. It has no shape and can be divisible. It produces a shadow so that the contrast between light and darkness is created. Once we know the elements, we write a high story of god. Everything converges into plasma or a fireball, i.e., fire. Fire decides birth, death, and time.

- It travels everywhere.
- It is unique: one God in three forms.
- It has three colors (red, yellow, and blue).
- It has three frequencies (low, medium, and high).
- It takes many forms (gods), colors, and temperatures.

Heat creates rain, clouds in the sky, and heaven, and it divides heaven and Earth. Light moves in a straight line. Heat moves in any direction. Initially, there was no light, but heat was present.

Imagine a strongly attractive hole in the Universe that eats up everything like a black hole. We do not know what happens to the objects after their absorption. However, after some time, the hole emits another or gives back to the donors. This is possible if another hole is rotating in the opposite direction. So these two black holes of opposite spins interact with each other, producing wave-like gravitation. In general, this friction produces heat waves. Now compare this with ancient rituals of producing fire where the heat of the fire was increased by adding inflammatory objects like alcohol. Fire absorbs energy, and a big person called Purusha emerges from the fire and gives back a source of life or fertilization or cell division. This product of the ritual breaks up itself into several pieces, and different forms of creatures can be found. (In the Indian Epic *Ramayana*, Agni initiated the birth process of the three queens by splitting itself into four portions of a special soup produced from the rituals.) In *Rigveda*, Agni is described as a Purusha having four horns. Agni can be split into many pieces, i.e., into an infinite number of pieces without losing their originality.

Heat is the only source of life and is unique. However, it is supported by an inflammable substance like oil. The interaction between the oil and heat produces fire or fire-oil produces heat (or light). These are Purusha and Prakriti, who always interact always. One is static, and the other is dynamic. After a while, the static object becomes dynamic, and the dynamic object becomes static. One produces another. However, when the dynamic becomes static, it gets split into several pieces to maintain the equilibrium. This is described in the mythological story of Shiva and his wife Parvati (or Sati), the sacrifice of Sati.

King Daksha excited/humiliated Shiva, and his wife Sati died. Shiva and Sati represented static and dynamic objects or Purusha and Prakriti in Chapter 4. After the sudden death of Sati, this order is reversed as Shiva became very active and started rotating the corpse of Sati. This is a pure nonequilibrium state, and the preserver, Vishnu, divided the body of Sati into pieces, and then equilibrium was regained. This transition from nonequilibrium to equilibrium has been followed in Sankhya, which originally came from the 10th Mandal of *Rigveda* where Purusha divides himself into many parts and creates life in different forms. So one unique fire divides itself into various parts to maintain equilibrium by many means—i.e., multipaths, multienergy levels, or multiple reflections. When heat energy is absorbed into a body, multiple reactions take place. In this way, the source of heat is divided into many paths, i.e., small sources. A heavy atom has several orbits compared to a light atom. By rotating a body in the form of a spiral, one can control the incident energy. Light can pass through a transparent object, but light can be captured if several mirrors can be positioned so that the same light can undergo multiple reflections and rotate in an orbit. This can make a core of light that cannot escape from the vortex. At the end of the cone, it rotates in the opposite direction and expands. This contraction and expansion continue forever.

This ancient concept is used to describe a black hole, which is a massive body that absorbs everything like a vortex formed in the sky. Gravitational attraction is the only cause that overcomes all repulsive forces and makes a unified system which is a unified god or the mouth of God who eats everything and digests them like fire. The attractive force works like heat, which is used to break up a large object into many pieces as the fire is divided into pieces. For example, heat divides still water into small particles of water which go up to form clouds, as if it has many heads, hands, or layers. After condensation, the water returns to the Earth as raindrops which are described as the many legs of Purusha. You need someone to excite this process. This can be a disorder that spontaneously breaks the symmetry or equilibrium, which is like Ashura or demons. Goddess Kali tries to regain equilibrium. She first breaks up the body of Vishnu into many parts. So a Purusha is fragmented and fights the Ashura. At the end of the process, the many facets of Purusha become one unique Purusha. In Sankhya philosophy, the breakup is described logically in Chapter 4.

Veda

Veda is the first light from heat. There are two types of knowledge of : *Para* and *Apara*.

Para vidya is supreme knowledge about the Universe or creation. This universal knowledge is self-manifestation that is created by itself or by its wish. No one created him. Omnipresent, he works as he wishes perpetually. He was self-birthed and self-working. When he was born, he was conserved by an object that was nonexistent, like heat, which means the world was cold before creation. He had only potential energy (no kinetic energy since no force was acting on him). All field lines are equal in all directions (monopole).

He is like a point object (dimensionless) that is surrounded by a field (his field) like a photon and electric (or magnetic) field vector. This configuration is Sunya, zero (not a circle or sphere), but symmetric, resembling a circle or sphere. Atman was neutral. No positive or negative things were associated: no polarity, no parity, no mass, no zero mass, everything undefined. Before or at the time of creation, there was no difference between excess and deficient. Both sufficient and insufficient did not exist. Neither death nor immortality. Neither air nor any other elements. There was no difference between day and night since the Sun was not formed. But Atman who is self-born had a signature of life and was working internally with his consort. Maya, or intellect, is an undivided part of him. Atman or Brahma was neither true nor false but was the smallest, indivisible part of fire that was ignited by his own constant Maya (like a point that glows, a photon covered by a magnetic or electric field). Brahma is inactive, but his consort Maya or intellect wished to work. Wish is compared to the force that can overcome, Inertia. Brahma, together with Maya, intensified the wish. This is like the inflation of space. Brahma wished to get started moving upward, i.e., adding a force. E_p is converted into E_k. This can be possible if acceleration is added with a greater and greater potential difference (wish), i.e., by adding a force. So, E_P to E_K is related through Agni = Acceleration. Imagine that Brahma and Maya are together, not separated, and that, due to inertia, no E_K is gained. By adding acceleration g or a to Maya (m), the particle velocity increases. So it gets separated from the (fixed)initial point; however, the potential energy (Brahma) is increased. To move the particle, Brahman creates space. This space is sky-like. This is created to accommodate Him as He grows up (or expands). As He created the sky (heaven) to accommodate the force (or kinetic energy) that He wishes to grow, He becomes associated with the increased velocity to support His wish to grow. This acceleration is called Agni (fire), which helps Him to grow. This growth can be in different directions in space, which can be infinite in time.

Starting from the water molecule (H_2O), the smallest object, He became Agni to carry forward everything. Agni drags particles and is responsible for all work. If we consider the creation of a star or the Sun, we can explain that it starts from a singular point like a black hole containing enormous energy. However, after the so-called big bang, particles move away or accelerate. This process continues forever, and time starts from the point when he wishes to expand and fill up the space He creates. He is the cause of gravitation or potential energy corresponding to gravitation.

After the 'sky' is created, all rays become aligned in the sky, and white light is created. This is called radiance—the Sun, which is covered with its heat, rotates around itself, and forms a space that is ellipsoid for its motion. All forces are generated from the inner part of this cosmic egg. He is called Purusha. He started from the atom, a singular point, and creates a big atom since he expands himself; however, He exists within everybody as a point object. When he becomes expanded, the Purusha is represented with as many heads, eyes, and legs as possible. The same original particle (photon) is distributed in many states. The last energy/potential energy is divided into thousands of Purusha who represent the Purusha. If we consider him as an ultimate oscillator who produces sound and creates cosmic harmony, then this sound needs space to propagate. Due to the constant oscillations of the Sun, sound waves are produced that can propagate over an infinite distance. This is like our heart oscillator, which is a signature of life.

Although He influences all waves and time in the world, He (the Sun) remains above all of us. He is above past, present, and future. However, He comes down to Earth as fire. He spends his energy creating clouds, lightning, and rain. He produces all cycles in nature, a time cycle. He created the cycle/Universe, watched his creation, and wanted to have a part of the creation. He wanted to create his form (image). He has thousands of eyes, as the Sun has many rays. He represents the Sun. He lives as air, sky, rain/water, Gods, and everything.

The formation of space attracts other fields, which increase density and, again, increases fluctuations. This process increases the speed of particles, which increases temperature that, again, increases the curvature of space. An increase in the curvature of space finally collapses. Space holds all waves/energy within itself. Finally, it collapses, giving light (fire). This is called unification.

He is very bright, and his shape is like an egg (zero). He moves in a space that is egg-like (Brahmanda). He lives in an infinite egg (Pura), he is the God of that space, and he is called Purusha. He sacrificed himself in his fire. He splits himself into many pieces and

FIGURE 2.3
The creator or the first person in history bears similarities in different cultures described as a bird-like character. (Left) A sculpture showing person sitting on a big bird. (right) A Sumerian mural shows a bearded person having two wings.

shapes. He is the first sacrifice. Brahma emerges from the cosmic egg. However, before creation, there was nothing but darkness and zero temperature everywhere. Then the water came into existence (plasma everywhere as the temperature rises). From that spray, a fiery golden egg consists of two parts of the shell: heaven, and Earth. Out of this comes Brahman, creator of the Universe, with the luminaries (the Sun and Moon) as his eyes. Mahasunya consists of many zeroes together in a big circle. At zero temperature before creation, the size of Mahasunya is infinite. With the increase in temperature, zeros split and got separated as time passed. However, at a much later time, the creation was described in *Brahmanda Purana* where Brahma was created from the body of Vishnu when He woke up from a deep, cosmic sleep. This is an introduction to the subject: Details of energy are given in the next chapter (Chapter 3).

Using the power of His mind, He made the Universe and produced four sons (four heads or directions) and then another ten men (all directions). Brahma divided himself into two and from His left half emerged a beautiful woman. Brahma's desire for His daughter was so powerful that He sprouted three extra heads (one on each side and one behind) so that He could always look at her Sun/Moon/Earth watched by Him or Sun divides into dawn, noon, and dusk to look at Earth). Discomforted by His lustful stares, His daughter rose to the sky (Ganges), so Brahma grew a fifth head that looked heavenward. The daughter fled, adopting various female animal forms as she ran (goose, mare, cow, doe, etc.), but all her disguises were in vain. Her father turned into the male counterpart of each animal she became and forced Himself on her to create all animal species on Earth. The daughter became known as 'Shatarup(one with a thousand forms). Again, we see thousands of forms like Indra in *Rigveda* having thousands of eyes (starry night or Sun rays or thousands of flames in a fire). We would like to change the idea of many to one or one consisting of three goddesses. They are Ila (on Earth), Saraswati (in the upper atmosphere), and Bharati (in heaven) like Vishnu, Shiva, and Brahma.

FIGURE 2.4
A statue of a god-like person has six main hands and several smaller hands which point in different directions. Three smaller heads can be seen on top of his head. Many hands and several heads from a body show the power of the Sun in sending energy (light) in all directions.

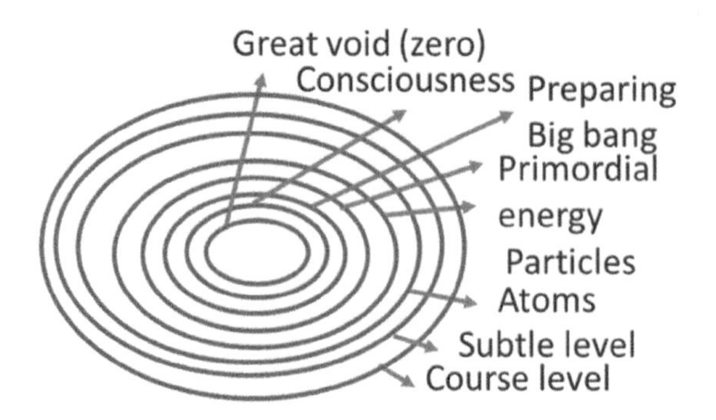

FIGURE 2.5
A holistic model of the Universe consists of various layers separated by a vacuum or empty space. A set of eight concentric circles representing different stages of the creation. Starting from the smallest circle as the beginning of the creation a macroscopic level can be achieved at the largest circle. The model of the universe includes illusion or uncertainty which gives false information. Truth must overcome this.

FIGURE 2.6 (a)
Three prime gods form one character, one god that has all three characters. A god with three heads, six hands, and two legs. Each hand is carrying a weapon such as a wheel, a conch, a water pot, a garland, a musical instrument, and a trident. However, they represent one god with three characters.

FIGURE 2.6 (b)
The Greek goddess Hecate portrayed in triplicate.

Modified Concept of the Supreme Creator: The Bramhan, Analysis of *Upanishads*, Liberation

When I rub two hard stones, at first it does not create any heat or fire; then I try to rub them periodically at a high frequency, and after several tries, I find that I am moving one stone onto another in a circular way. I find sparks coming from the contact point, and the direction of the spark is perpendicular or at a tangent to the circular path. The many sparks resemble fire with many heads. These sparks go out of the wheel and do not return to their origin like hot particles. But cold particles can revolve on and on to attain the

FIGURE 2.7
Holy trinity is celebrated in Christianity. Father, the holy spirit, Jesus, Mary, Joseph, and other characters are seen in the picture.

threshold energy. This is called 'liberation' (of Atman). There are two types of Atman: one confined and the other liberated. Confined Atman is like heat that gives a cycle or periodic oscillations. It is manifested as light that ends the cycle, which is called liberation or liberated Atman. When our body dies, the heat of the body leaves us as photons at a higher altitude. If this heat is enormous or sufficient, then the particles can go to a very high altitude as high as the clouds. After the collection of particles, this is dissociated into lightning. Heat is converted into light, which is an example of liberated Atman. If the heat is not sufficient, the particles cannot reach a high altitude and come back to Earth and will continue with the cycle. Now imagine I am Atman, and my representation is invisible heat and light, but I cannot be seen unless I interact with matter in the form of absorption and reflection. So I am converted into heat. I am like the Sun, a ball of fire that emits light that must be controlled or converted into heat, so I create my consort, which can absorb or reflect the light I emit. My consort creates shadow. Any observer on Earth can see my light and the shadow. This happens from the day of the solar eclipse and is fully or partially covered by the Moon. So now this shadow and light create a superposition of states. It is a duality. So I the Sun or (prime god) Shiva with my consort, the Moon (goddess Kali), produce Maya through the reflection and absorption of my light and shadow. Since, I, the Sun, create all kinds of heat on Earth, including the heat in human bodies, after death the released heat from the body must be liberated.

Summary

One of the most ancient of scriptures, The *Rigveda,* introduces many gods. We explained three cycles, namely Indra, Varuna, and Agni, as the three prime gods in this book. Finally, this book accepts the concept of one supreme god. She is the universal mother. The model of a universal god was developed further in the *Upanishads,* which influenced the idea of formless gods in many other cultures. The characteristics of gods are described in the next chapter (Chapter 3).

References

Vedas and Astronomy

Vedic astronomy and post-Vedic astronomy. Vedic astronomy is the astronomy of the Vedic period, that is, the astronomy found in the Vedic Samhitas and Brahmanas and allied literature. The principal avocation of the people in Vedic times was the performance of the Vedic sacrifices at the times prescribed by Shastras. It was therefore necessary to have accurate knowledge of the science of time so that the times prescribed for performing various Vedic sacrifices could be correctly predicted well in advance. *Upanishads, Brahmanas, Smritis,* and *Puranas*—which sometimes claim to be as old as Vedas—also include stories and references from the post-Vedic era. It is believed that these texts have evolved over centuries, with some of the last additions being done as late as just 2,300 years ago (around the time of Chandragupta Maurya). To date, different authors attribute different timelines to the Vedas, mainly based on the reference to stars and constellations, mentioned in different verses. The estimated dates vary from as old as 6,200 years (Orion being the first Nakshatra) to 4,700 years (Krittika at the vernal equinox point) to 3,400 years (Ashlesha at the summer solstice). A verse in *Rigveda* says, "The wheel (of time) formed in 12 spokes, revolves around the heavens, without wearing out. O Agni, on it there are 720 sons (that is days and nights)". So a year has 12 months or 360 days. Later in Taittereeya Samhita, there is a clear mention of a solar year of 365 days. The names of the 12 months as given in Samhitas are Madhu, Madhava, Shukra, Shuci, Nabhas, Nabhasya, Isha, Urjaa, Sahas, Sahasya, Tapas and Tapasya. Now, a lunar month is nearly 29.5 days and 12 lunar months make 354 days. There would be an extra intercalary month, or Adhika masa, called 'Samsarpa,' in some years to align the lunar months and the solar year. In *Rigveda,* it is stated that the god Varuna charted a broad path for the Sun in the sky. This refers to the ecliptic, which is the path of the apparent motion of the Sun around the Earth in the sky against the stellar background. It is inclined to the celestial equator, which is a large circle in the sky in the plane of Earth's equator.

Creation Stories Derived from the Rigveda

I do not know whether *Rigveda* is composed of old mythological stories of Gods, but we know that many, many stories are created on the basis of *Rigveda,* which says that the world was created from Nothing by a female creator Aditi, who gave birth to Daksha, who in turn creates Aditi, and a cycle of creation begins. An exchange or duality can be imagined, although the cycle is a completely timeless process of the god. To describe four fundamental elements, three main gods are described in *Rigveda*: Indra (god of rain and storms, a cycle of creation like Brahma), Varuna (god of the sky and celestial water, like Vishnu lying on the bed of water and creating the curved space), and Agni (god of fire or destruction, like Shiva, who transforms energy to matter and creates even rain by heating water and recycling it). Three cycles—Agni, Indra, and Varuna—were born from an earlier creator Purusha, a male primordial giant who released Indra and Agni from his mouth and Earth, as described in other cultures. Purusha had a thousand heads, a thousand arms, and a thousand legs. His three-quarters rose upward (like the three-quarters of the Earth that is water) and one-quarter remains here (like the one-quarter of solid Earth). From this one-quarter of Earth, he spreads out in all directions. People of different castes are created from his body: Brahmin from the mouth) warriors from the arms, traders from the legs, and lower-caste workers from his feet.

Creation of Sun and Planets

It was imagined that the Sun was created from cosmic dust, which was seen as a cloud and revolving around the center of the Universe as well as its center of the gravitational field, i.e., Vishnu. The gravitational field of the Sun is Indra, which holds all the satellites together. However, other forces, like electromagnetic forces, hurried to disturb the systems but failed, being treated as demons.

Creation Stories from Ocean (gods vs. demons)

This describes a process of creating rain. It seems that clouds were blocking the sunlight, and water was not formed on Earth. Indra used his weapon to break up the cloud. Vishnu, or Sun, absorbed water for eight months and gave rain for four months, which cleaned up the Earth. Since Indra could not get rain through prayer, Sri Krishna Vishnu had more of a cosmic picture.

Three Cycles of the Vedic Gods

The gods of three cycles are described in the *Rigveda* (10th Mandal with Visha Karma and Prajapati [Brahma]).

Indra

This god with many heads/many hands/many legs liked to kill demons. He destroyed clouds and created a fire in the sky and then rain with his weapons like thunder. He is compared to the Greek god Zeus, the god of the sky. Indra is Sun and fire. Indra is the sky, which has thousands of eyes (the stars). Many stars together or the many rays of the Sun or its many flames are equivalent to many eyes, horns, legs, and hands—i.e., many atoms/Atmans. Sri Krishna opposed Indra-related prayer and saved people from the heavy rain created by Indra. Vishnu replaced Indra.

Varuna

Varuna is like the god of water/sea etc. but of the Earth (instead of Indra, Lord of Rain). He has been found to be similar to the greek God Ouranas as the universal god of the sky, like the Jewish Jehova or Iranian Ahura-Mazda or the Hellenic Poseidon. More precisely, he is the god of the night sky but is also the Sun god.

Vayu

Vayu moves in the upper layer of the Earth, which resembles the breathing of the sky. It carries the energy of the Sun and transfers the heat from one place to another. By changing the velocity of the air, the rate of transfer of heat energy can be controlled. This is very important to control the temperature of the Earth. Heat is transmitted from the surface of the water reservoir to the cloud through water vapor.

Agni: God of the Sun

Agni is the prime god of Veda after Indra; he is the mouth of the gods. He is everything in a Jagga from priest to sacrifice to God—one God for all purposes. He is the Son of Energy of Force, which is created through interactions. Agni is a purifier to water through a cycle. He is the Sun god at night. Agni and the Sun both give light and heat. They give life. Agni has many heads of flames. He lives in three places: heaven, the clouds, and Earth. People got frustrated with the 'God' description and wanted logic. Moon was born from the mind of Purusha, the Sun from eyes (cycles). From his navel, the middle realm of space arose, and the sky evolved from his head. Earth came from his feet and the quarters of the sky from his ear.

Daksha (Shiva-Uma-Vishnu)

This creates a ray of the Sun rotating around the center of the Universe, the Earth is rotating around the Sun, and, most importantly, the Moon is rotating around the Earth in 27⅓ days, which connects Sati, representing Vishnu (gravitation field), to her husband Shiva, representing Sun.

The *Upanishads*

Isha Upanishad

Brahma can attain the highest speed. Air or wind carries the heat. Vidya is the knowledge for unification, and Avidya is for the many of multiplicity. So Vidhya and Avidya correspond to the attraction and repulsion associated with contraction and expansion. Since Maya is a repulsive force, it tries to compensate attractive force toward the center and throws different bodies. This is like centripetal and centrifugal forces, which can coexist in a rotating body. An increase in repulsive force means creating many orbits. It is interesting to know that a black hole has an exchange attractive force but that, still, it emits some radiation as a result of repulsive forces. The big bang is a result of repulsive force.

Kena Upanishad

Atman is the complete body consisting of the sense, mind, and intellect. It is like light that is focused by the eye into an image although it cannot see photons. Agni, which is visible, is not considered Brahma. Upanishad did not admit Air, Agni, or the Sun as Brahma or the main cause of all effects.

For example, the Sun is covered by a golden plate, and Agni is visible. However, the Sun is described asphases or gods, e.g., Pushan, Ekarshi, Yama, and Prajapati. In the *Kena Upanishad*, it is compared with lightning, which lasts for a very short time. So the main cause for all effects is heat, which gives concentration of the mind. It serves the purpose of unifying body and mind and the senses.

Katha Upanishad

It describes three types of Agni who live in heaven, sky, and the land, like the Sun, lightning, and fire on Earth. Agni can attract others and divide them into parts. The word 'AUM' is created and is composed of contractions and expansions—two opposite forces working together. So the inside as well as the outside our bodies are caused by attractive and repulsive forces. They are still motionless, however, and can travel a long distance because they can spread out by the tunneling phenomenon. Senses or Indriya are moving in an outward direction as if repulsive forces are acting on them. But we need attractive forces. So we have to change the direction from outward to inward so that we can see the origin. Here Agni is considered as Atman, who makes our Sun rise and set since the Sun has no active role and follows the laws of Atman. Atman, like Agni, is indivisible and cannot be multiplied. Purely this is the gravitational force.

Prahsna Upanishad

Dualism is a cause of creation. Prajapati is the creator whose role is to create, extend/increase, and divide (multiply). He created heat, followed by Mithun (love) or duality. Mithun means unification of Prana or living force and Rayi or matter. The Sun and the Moon are described as Prana and Rayi. So the integration between the Sun and the Moon is gravitation.

Mundaka Upanishad

Purusha is described as a bird with large wings. Agni arises from Absolute Purusha. Agni has five types: (1) the Sun in the sky, (2) clouds that give us water as rain, (3) Earth that gives all plants and trees, (4) Purusha thst gives birth, and (5) the wife who gives us life. However, Brahman and Atman are different from Agni defined in other parts of the *Upanishad*.

Mandukya Upanishad

Atman has four legs, like the four legs of a cow, although Atman is indivisible. It has a gross and three subtle parts: (1) Baiswanar, (2) Taijas, (3) Pragna, and (4) Turiya. In other words, Atman has four dimensions A, U, M, and no dimension. A means spreading out or expansion, U means contraction, and M controls the expansion and contraction; both repulsive and attractive forces die out at M. Chanting AUM resembles three tones—A, U, and M—but there is the fourth one above AUM that connects Atman to absolute Atman. It is purely nondualistic. We can think this fourth dimension is like time independent from the three others.

Taitriya Upanishad

There are four dimensions of Atman: Bhu, Bhula, and Swar, which correspond to Earth, sky, and heaven, and the fourth one, Maha, which is like the Sun and gives us light and heat. 'Maha' means great or infinitely large. These four things also correspond to the different parts of Purusha like two hands, two legs, one head, and the middle part of our body, where Atman lives and which also expands like an egg. The sky is produced from Atman. Air flows from the sky, Agni from the air, water from water, and Earth is created from water. Earth creates plants and food, and, finally, Purusha is created from food or Atman, as an egg is created in the body of the mother. Purusha is a birdlike creature that has one head on the shoulder, its left hand is like a left wing, its right hand is like a right wing, its main part of the body is where the heart or Atman lies, and it has a tail below the navel part. We are created by our father who is a replica of our father. In the mother's womb, we exist as Atman inside a shell full of nutrition that is separate from the physical body. However, it has a shape like Purusha, which is made from Vayu of fire types. Air (Prana) is his head, Air (Vayu) is his left wing, Air (Apan) is his right wing, Air (Shaman) is the middle part of Atman, and Earth brings stability as the tail of the bird Purusha. So the egg and bird model is the model of the Universe. The Universe is a combination of the bird-like Purusha and the egg-like Prakriti, and they are interchangeable like the chicken-and-egg problem. The egg gives birth to a bird, and the bird lays an egg and observes the egg until it gives birth to a bird. Many eggs can give birth to many birds. Likewise, many birds can lay a lot of eggs. So one Purusha can divide itself into many Purusha or Prakriti. This originates from the subject light (bird)-atom (egg) interactions. Like wings, light has polarity, electric, and magnetic fields that interact with the shell (egg) of matter or atoms.

Oitemya Upanishad

The egg model of matter or Earth has been thoroughly described in this great *Upanishad* derived from the *Rigveda* directly like multiple worlds were created by one unique Brahma, or Sun. He created four worlds: Ambha Lok which holds water like clouds; Marich Lok, like the sky; Moro Lok, like Earth where everybody is mortal; and Apa Lok, the world of water under the Earth. Brahma created a form of Purusha from water. Here Purusha is like an egg. Brahma tried to give life to an egg. Due to his wish, the egg was manifested as a mouth and then as the organ for producing a sound that gives birth to the first God, i.e., Agni. Then the organ related to respiration and smell was created and is assigned to the god Vayu. Then our eyeballs were created, and from the vision-sensing organs, the god Sun was created. Then the ears are formed, followed by the creation of the god of all directions. After the creation of touch-sensing organs, the heart is formed, followed by internal structures in the body. Then our navel is created, which manifested the god Apan from which Apan, the god of death, is manifested. Finally, the sex organs are formed, followed by their creator Prajapati.

Swetaswatar Upanishad

This describes a supreme god who is the cause of creation, preservation, and destruction, but he is beyond all these characteristics. He creates all kinds of dualities. He is unique and can also take on three characteristics. He has an endless number of heads, eyes, hands, and legs. This infinite Purusha lives everywhere, but in a very limited space. He is described as the prime God, Shiva, or Rudra representing the destroyer. He can create fear among us and raise devotion to the followers. The power or energy of God is Yoni (opposite to the Lingam). He is described as a butterfly or a blue bee or a bird having two wings and a body (head) at the center. After all, He represents Agni who lives in the upper layer of the sky and is equally powerful to the Sun. This is the power of electricity (lightning) that belongs to the cloud in the sky.

Chandagya Upanishad

Purusha lives in the eye since the eye reflects the world: a holographic Universe, white (Sattwa) + blue/black (Tama), etc. The white and blue parts of the eye correspond to the Rig and Sum Veda. This means focusing to a point. If you add water to the eyes, it will be deposited on the two sides of the eyes, which gives a dualistic view. Tears come out of the eye from two ends and do not return to the eyes. Liberation comes from excess tangential velocity, which is like light emission and which does not return to the periodic trajectory. The liberated heat produced from the oscillator present in the body changes its lifetime or dependency on the strength of the liberated Atman. He goes from heat to half of a day, then from a day to half a month, from that to half a year, and then completes the cycle of a year of the Sun. From the Sun it goes to the Moon through the light and finally to the world of lightning (inside clouds). Then a nonhuman escorts the Atman to Brahma with a chariot known as 'Devjan'. On this path, there is no return.

3

Life Begins on Earth—Creation of the Devices, Deities, and the Concept of God: Creation of Cycles

Slaying of Demons Madhu AND Kaitava by Lord Vishnu

At the beginning of the Universe, there was an ocean all around. Vishnu was floating in the middle of the ocean, under the influence of the goddess of Sleep (Yoga Nidra) and a lotus stalk and a flower grew out of his navel. The God of creation 'Brahma' was born from that lotus flower, and He was in deep meditation at the time of His birth. At the same time, upon reciting Vedas, ear wax flowed out from both ears of Vishnu, and two demons were born out of the ear wax. These two demons came to be known as Madhu and Kaitava. Both of them performed great penance for thousands of years, without taking any food and drink but constantly chanting the sound of mantras that they heard before. The goddess Shakti was pleased with their devotion, appeared before the two demons (Madhu and Kaitava), and granted them the boon that their death would come to them only when they desired it. With the knowledge of their immense strength, both the demons became arrogant. They attacked Brahma (the God of creation) and stole away the four *Vedas* from Him. Brahma became furious but was helpless around Madhu and Kaitava. Hence, Lord Brahma started praying in great consternation to Lord Vishnu to seek His protection.

Lord Vishnu, however, was in a deep sleep, under the influence of Yoga Nidra and did not wake up despite the best efforts from Brahma. Brahma then said, "If Vishnu is not waking up, it must be the work of the Goddess of sleep. I must firstly placate the Supreme Goddess so that Lord Vishnu would wake up and save me". Thus Brahma started praying to The Goddess. Hearing the prayer the goddess of sleep took pity on Brahma's plight and left Vishnu's body. Then Lord Vishnu was freed from the influence of Maya, woke up yawning widely, and saw the mighty demons threatening all creation.

Lord Vishnu fought against the demons Madhu and Kaitava using his arms as weapons. The battle continued for many (5,000) years. Lord Vishnu was surprised at his inability to destroy the two demons, Madhu and Kaitava. Vishnu then divined that these two demons had obtained the boon of invincibility from the Goddess of Shakti. While Lord Vishnu was still battling the demons, He started to pray to The Goddess Mahamaya (the great illusion) to show a way to slay these demons. The Goddess appeared in heaven above the battleground. He beseeched The Goddess to show a way to overcome these demons, who have become strong as a consequence of The Goddess's boon and who cannot be killed if they do not desire it. The Goddess Mahamaya suggested that these asuras must be defeated by tricks. Illusion was the right weapon to use against these warriors. She cast a

DOI: 10.1201/9781003304814-4

FIGURE 3.1
Hindu cosmogony: Brahma, seated on a lotus, rising out of Vishnu, who lies asleep on Ananta, the five-headed serpent, while Lakhsmi, his wife, sits at his feet.

look mixed with false affection on the two demons, causing their baser instincts to awaken. The two demons fell under the spell of The Goddess. Since Vishnu could not defeat them, he offered them a false boon. In return, the demons offered Vishnu a boon. Vishnu asked them to die in his hands.

Too late, the two demons (Madhu and Kaitava) perceived that they had been tricked. However, there was nothing to be done now. They had promised to grant a boon and were bound by their promise. They just had one stipulation to make. They did not want to die on water. Lord Vishnu said, "So be it!" He then took up His Vishwaroopa (a boundless form, immeasurably immense). He then placed the two demons on His massive thighs and beheaded them with His Sudarshana Chakra (a special disk-like device). From the body fat of the dead demons, the land was formed. Since the land was formed from an unclean source, the soil is not considered to be fit for consumption. Thus Lord Vishnu, in His manifestation as Hayagriva killed them and retrieved the Vedas. The bodies of Madhu and Kaitava were disintegrated into two times of six—which is 12 pieces (two heads, two torsos, four arms, and four legs). These are considered to represent the 12 seismic plates of the Earth. This story of creation can be applied to the Sankhya philosophy.

Creation Stories

Sumer

There are lots of similarities between Vedic gods and Sumerian gods in terms of names and activities. It seems the word 'Sumer' comes from the pole of the Earth, which is Shivalinga (or Brahma). Sumerians also described the birth of god from the deep ocean (water). Apsû is water ('Apas' in Sanskrit means water), and tiâmat is deep ('Tamas' in Sanskrit means darkness). Tiamat plots to destroy her divine offspring. The eldest chief in Sumer 'Mummu' means Manu in Sanskrit. Similarly, Laghma is Laxmi, Anshar is Akash (the host of heaven),

(a)

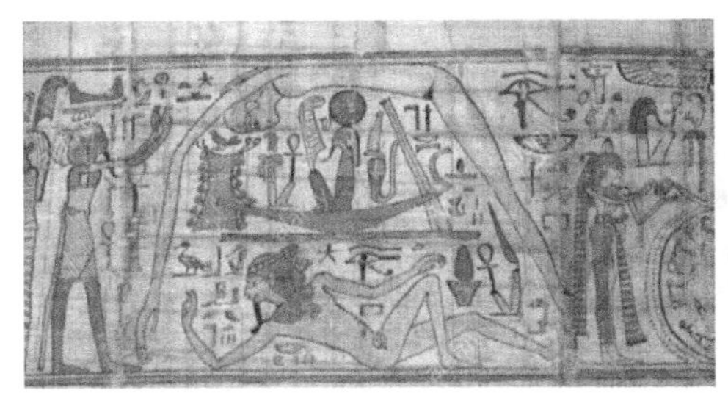

(b)

FIGURE 3.2
Creation described in (a) Sumerian and (b) Egyptian cultures.

Kishar (the host of the Earth) is Krishna, and Shar (the host of the ocean) is Sagar or host of Sat, i.e., Sangshar, in Sanskrit. In the emblem of Sumer, we see the Sun with wings, which is very similar to Brahma's Maya. The five circles correspond to five elements, like Shiva having five heads. People with long beards symbolize something covering the face of the Sun, i.e., Maya. Lion is the power of the Sun but granted with wings, i.e., Prakriti (Maya). The four-headed Brahma-like feature is found in Sumerian artifacts. Beyond wings, snakes correspond to folding or spreading, which is nothing but Maya. Finally, the Durga goddess is found in Sumer with ten hands or ten forces. Concentrated at a point, the carrier of Durga is Lion (a masculine form of power, Sun) is found in Sumer. Both the Sumerian and Indian cultures are based on Sun, which shows solar rays through wings in many places. Dualism is found in ancient Sumerian civilization from the creation of man and woman. 'Adam' means single spouse. All creatures are created in pairs (man–woman and light–darkness). The Assyrian equivalent 'Namassu' means God, which can move or creep like a snake. This is like the spreading out of a snake. 'Namassu' in Hebrew means Ramas, to creep. It seems the word 'Rama' came from an object that can creep like Vishnu. We see Vishnu sleeping on the bed of a snake and then spread. Hence the creation story is very similar to Sumer and Vedic civilizations.

Egypt

Like Brahma's four faces, Sumer and Egypt show four elemental forces or elements. God was lying on the bed of water, so Nun is similar to Narayan. Egypt was floating on four pillars. Ra is the Sun god, whose movement trajectory makes a dome atmosphere that moves from east to west. The creation story is like Vishnumaya. Clouds act as a demon cover, created by the Sun or Vishnu through water vapor, and light cannot reach Earth.

Heat is released from earth or water. The cloud is cracked by an electric spark (Agni) and raindrops. Earth is like a mother cow (Go-mata). Like Krishna, the Egyptian God is protecting Earth from Indra, the god of rain. The Sun god is sitting on a coiled snake that is like a cloud. Even God and demons are pulling at the head and tail of the snake, and the center remains stable. Three gods hold heaven. A pyramid is like a flare of fire pointing to heaven. The head of the pharaoh is covered by a special cap, which is a combination of fire and water like the Sun and Moon together. Osiris is the god of death, and Ra is like Rudra (Shiva). An antigod (Seth) is an Apophis snake. This shows the duality of the Sun and Moon. Stories described in Sumerian/Egyptian or Greek mythology and even in the early *Rigveda* about creation by god should be summarized into rules, mathematical formulae, or equations in order to be applied in modern science, physics, chemistry, biology, etc.

The Puranas: Vishnu Maya

According to *Sri Brahma Samhita*, Krishna is the supreme entity who is manifested as the all-pervading Paramatma, Isvara, or Vishnu. He lives in Gokula, which is a sphere (Goloka) and the center of the thousand petals of a lotus. Krishna or Govinda is the prime cause of all causes. He is surrounded by the halo of Brahman or an indistinguishable entity. The lotus is like a hexagon with sixfold divisions. It is like a diamond or the Sun. Sri Radhika, consort of Sri Krishna, lives in the petals of the lotus with other gopis. Maya of Krishna, Rama The Goddess, is regulatory of all entities. First, Krishna creates a divine halo, Sambhu, the masculine symbol. Krishna wishes these two, Shiva and Shakti, which are predominantly active potency (cause) and predominated active potency (effect), unite (a causal principle). This is the wish of Mahavishnu, who possesses thousands of heads, eyes, hands, and legs and creates thousands of souls. This is Narayan who lies on the water bed in a state of divine sleep. The spiritual seeds of attraction (sankarshan) exist in his bones on his skin like golden sperm covered with the five great elements. These pores are vortex-like.

FIGURE 3.3
After a long fight, Vishnu managed to kill the demons Madhu and Kaitava by separating their heads from the bodies using his unique wheel-like device, the Sudarshana Chakra and created soil.

The rays or jivas who had lain dormant during the cataclysm are awakened. Shiva is pulling up these jivas or rays from the body of Vishnu. They are elementary particles and are not perishable. They are atomic particles: photons. The Sun god is spreading his rays as described in the *Rigveda*. Now he creates Vishnu from his left limb, Brahma from his right limb, and Sambhu (halo) from the space between his eyebrows. After creating these three gods, Maha-Vishnu prefers to consort with the goddess Yoganidra or Mahamaya. Now Brahma rises from the naval of Vishnu with four faces (four types of knowledge). After his birth, Brahma finds darkness around him. With the help of The Goddess Saraswati, Brahma rises. He absorbs sound and has a rebirth.

This model can describe the formation of the Sun from gas/plasma as follows. A transition from a flat space to a curved space (magnetic belt) can be described by the rise of Vishnu. From some scattered gas or ions, a belt is created due to rotation and gravity (angular momentum). This ball is squeezed into a small volume. Contraction can be created by a gravitational pull. Anyway, when charged particles are aligned or ordered in a smaller volume, a magnetic field or magnetic lines of force are created. A gap is created, which is darkness and holds the center or Sun like a mother holds the child, Sun. Magnetic lines of force can create a cover/shield around the bright sphere outside the gap, which is a halo. This cover is Tamas. Sun is active internally due to nuclear reactions. With the cover, heat cannot go out, and light is not produced due to the lower temperature. Due to the reflection of the heat from the shield, the temperature rises and eventually creates photons that penetrate the shield. All particles that were originally aligned on the surface of the Sun (along the magnetic lines of force) are inactive. Now photons are more perpendicular to the surface and spread out in all directions, helping to stabilize the magnetic field. Both photons and the magnetic field lines represent Kali who emits from the body of the Sun (Vishnu). Magnetic poles are the faces of Brahma.

Among the creation stories, I think Vishnu Maya has the best description of the 'creation', as elaborated here. Sri Vishnu is the Purusha in absolute Brahman, power, and constancy and contains six characteristics. He is undistracted and lives with everybody. He is also Shiva (intellect energy of intellect). The goddesses are Shakti of four Buk; Bharati or Durga, within the Sun, deliver (Kali) Shakti. This force or Shakti places the Moon between Sun and Earth and stabilizes the Sun-Earth-Moon system. Vishnu or gravitational field is dispersed everywhere like consciousness. It is called Aditi in *Rigveda*. It is like inertial energy (dark energy, Tama-like). This Shakti works against the energy or force of gravitation of electricity that binds the masses or particles of opposite charges and forms a condensed state. We see evidence of the attractive force of Vishnu from the formation of the stars/planets and their motion. So this is the only field, which, like potential energy (cause), can attract thousands of stars (suns) toward the center of the Universe (or the center of the cosmic egg). It is called the consciousness of Brahman and creates a gravitational force for our Sun (called the consciousness of Indra). It also creates an attraction between the Earth and Moon and the gravity of Earth that attracts us. So the gravitational force field of Vishnu is the most universal field, which decays if the stars move away ($F\alpha 1/r^2$) and becomes strong with the inertia of stars, i.e., mass. These forces are called material-Shakti and sky-Shakti. Vishnu consciousness is a combination of these two Shaktis. $F = \frac{Gm_1m_2}{r^2}$ is a combination of the two Shaktis; hence the r dualism model can be revealed. The central electrical/magnetic energy of the Sun can be represented by the goddess Durga, and the energy surrounding the Sun is called 'Laxmi'. Earth has shields to cover it in the air or atmosphere, which contains heat to maintain the temperature for life. The back reflection of heat from the cover is Mahamaya, which applies to every physical process like the mother protecting her child.

In the first stage, Brahman was observing himself, which created the wish for creation. Brahman wanted to become many from one entity. He wanted to see his own characteristics, his own form, and own energy, i.e., self-form/self-characteristics/self-energy. This is dualism, and breaking dualism makes creation. This conjunction of Purusha with its own power, i.e., Shakti, makes the creation happen. Shakti lives within Him as condensed energy like Kundalini who is the representative of Vishnu but is separate from Purusha. Let's focus on Vishnu's Kriya (action) on Shakti (energy). Kriyashakti and Bhutashakti represent the wish of Vishnu and the material for creation.

Bhuti (E_P) is the cause of Shakti. The central part is like a point, a dimensionless object and the only point that lives with its own energy without separation. It is like a funnel, and the closer end has a singular point. It is also like a black hole or our mouth that is connected to the air coming from the internal part of the body. This is silent without any manifestation of work, the ultimate ground state of energy, i.e., Shiva. We need to produce sound by exciting these point objects. By its own wish, it divides itself into a state where the point is slightly off-centered, i.e., a very small gap between the center and the new position, i.e., duality. Once this happens, the Kriyashakti or kinetic energy starts working, and there is an exchange between the E_P and E_K. The unwinding of the system begins. To make a circle or orbit, three quantities are needed for the stability of the circular motion: (1) a fixed center (Satwa), (2) a circular speed or angular momentum (Rajas), and (3) constant acceleration or force toward the center (Tamas). So we create resonance frequency from the rotation (light emission). Many orbits can have different frequencies that can create mixed sounds/colors.

Kriyashakti and Bhutashakti are like heat or light and fire. Bhuta (fire) is surrounded by Kriya and is similar to the Sun surrounded by orbits. Heat surrounds fire in all directions and is controlled by the internal energy of fire. So this is like cause–effect (Laxmi–Sudarshana). Now creation is of two types: (1) pure creation (first light or vibration) and (2) purer creation (four stages of manifestation), implying a transition from one state to another. There are four layers: Basudev, Sankarshan, Pradumna, and Anirudha.

The first state of Vishnu Shakti, God Basudev itself, is absolute Prakriti, but it does not have the characteristics of Satwa/Raja/Tama. It is extremely pure Prakriti. The first nondistinction between Shakti and Saktiman is the Basudev theory. Both are indifferent to each other. Then Basudev wishes to split itself within itself. It is like before sunrise, Sun is spreading light although the Sun itself is not visible. This is the second stage of Shakti called 'Sankarshan'. This is like creating light, which gives red/yellow/blue light from the second state. In the third state, Pradumna is created where Prakriti is separated from Purusha. Three Gunas, i.e., Satwa/Raja/Tamas, are created. Finally, in the Anirudha stage, mass and energy are created with the help of time.

So Vishnumaya is the secret of rotation in a combination of its own inner Maya, i.e., potential energy, and other Mayas, i.e., kinetic energy, which can be divided into three parts: center, attraction toward the center, and rotational motion. This E_K produces a magnetic field or resonant sound that exists perpendicularly to the plane of rotation. Vishnummaya is the self (potential) energy that controls equilibrium. It improves the idea given in *Veda* around Indra and brings a revolution in thought, i.e., Maya that plays an active role in creation.

We see that the water cycle produces rain assisted by the Sun, which also creates a cycle of life. It connects two reservoirs through a cycle or a feedback loop. Water is heavier than air and cannot overcome gravity. So the Sun breaks up water stored in a reservoir into vapor, which can defy the attraction of Earth and overcome the inertia. It goes to a certain height and then collectively forms clouds, which also separate charges into positive and negative (closer to the ground). They recombine and produce light and sound. The water stored in the cloud can have a large mass and eventually come back to the ground as small particles of water, i.e., raindrops. Remember that the massive objects are Tamas, the cycle is Rajas, and the purified water is Satwa and that all three are parts of the cycle or an imaginary ring. The king of gods Indra used his weapon to break up the cloud to get water. Vedic priests used large fires to send water to a higher level that could produce rain. However, this kind of process did not work all the time, and people or scholars started thinking about a model that would bring water in order to at least know the logic (science) behind the cycle. It is not destruction (like Indra) but preservation that would be a better solution. Instead of killing animals through sacrifice, one should preserve them. The unified god of the sky, Vishnu, appears and could save the world from natural calamities. He could raise the Earth above the level of water. He could look after the Earth with three legs like the Sun watches from dawn to noon to sunset. Vishnu can split himself into three parts or properties Satwa (neutral), Rajas (active), and Tamas (inertia). Also, he can be in the fourth state, i.e., in the form of deep sleep. He is dreaming about the world. That means he is creating an image of the real world.

How can you combine a real thing with an imaginary quantity? This is possible by the addition of a phase, i.e., rotation ($x + iy$; $re^{i\theta}$). In other words, you are creating a cycle by rotation or a semicircle. Or this is like two circles moving in opposite directions. The part they touch is virtually static. This works like memory. Vishnu is preserving things by creating a static object. However, other local forces can disturb the cycle. As described in *Sri Sri Chandi*, some demons can disturb the system so that the static state cannot be formed or sustained. Vishnu wakes up from his dream with the help of Vishnumaya or The Goddess (Kali). Vishnu kills the demons with another cycle, i.e., his Sudarshan Chakra (a special wheel-like weapon).

In the model of the creation of the world appearing through Vishnu's dreams, all processes are described particularly as three characteristics, duality and nonduality. Starting from a flat space in his body lying on the water bed, a dome-like (spherical or semicircular) space is created. This sky is filled with stars that give light. When the lights from all stars are combined, then Brahma can rise as a representation of the Sun. So the Sun is formed from the energy of the stars. The curved space is created by contracting the space of Vishnu through gravitational attraction. The Sun is formed from gas particles in the Universe that are attracted to one another by Vishnu. This attractive force can compete with a repulsive force like electricity. However, gravitation works like love and defeats other destructive or repulsive forces and finally forms a material object. This can be preserved for a long time, which in turn protects the planets around the Sun.

The force of love was introduced by Vishnu. Brahman loves his Maya (its consort or its image) and creates a static object.

Unlike a single cycle, we now have multiple cycles. They can be arranged in different orbits around Krishna like Gopis (Woman). This is the model of the planetary system we use today from water cycle to cosmic cycle images. In *Rigveda*, Indra (representing the Sun) is not associated with Maya. The dualism concept was developed with Vishnu. Together with The Goddess or a consort, cycles can be made more robust (sustainable) than a single one.

Maya-Shakti: The Goddess

The most important thing is to bring The Goddess into the model. The Goddess is a signature of the mind. Mon or Mond is a Shakti and time-dependent part of static Brahman.

It is a change of phase, like the Moon shows every day/night. So the equilibrium picture is complete with a combination of the Sun, Earth, and the Moon as the mediator or connecting the Sun to the Earth. The Moon was present earlier on top of the head of Shiva in his meditating posture. A cycle of water (Ganga) was also shown on Shiva's head. But Lord Shiva was shown to be the dominant power and active. For Indra, his wife Laxmi or Sachithe Goddess was described briefly, but stories of a heroic nature were shown instead of her active role in the battle. Vishnumaya explains the modern concept of the Universe where he remains inactive and The Goddess is active. This creates a new subject on the active role of The Goddess, which excited Vishnu. So there is a Shakti, which lies in the heart of Lord Vishnu and which is active and separate from the physical body of Vishnu (the Sun) that gives Vishnu the power of becoming many from unity. Without this Shakti, Vishnu cannot divide himself or spread out, and then the Universe cannot be formed. This is something like the nuclear reaction that takes place inside the Sun, which yields light and heat from the Sun. However, some kind of forces always oppose the creation process. They are described as demons like Bitra in *Rigveda* (killed by Indra), Madu-Kaitava in *Sri Sri Chandi* (killed by Vishnu), Mahishaur killed by Durga, and Chanda-Munda (or Sumbha-Nishumbha) killed by the goddess Kali.

These processes can be formulated as mass–energy conversions. M_t (full mass or matter) = m (inactive or inertia) + m_{es} where these two masses interact with each other. The stored energy e_s is a part of the big energy (E_t) used to activate Vishnu. However, e_s is much less than E_t ($E_t \gg e_s$) through Yoga Nidra, or cosmic sleep. Vishnu converts matter into energy, which is used to activate an inactive or inertial body. With the relationship $M = m * t * e_s (c^{-2})$, God is making a conjunction of his spiritual potency with his inactive, nonspiritual potency. So this is also a kind of purification of energy like rainfall or a cycle that purifies water. Here M (Paramatma) (Vishnu) connects to Jiva Atman (Brahma) where e_s is used for the activation of Vishnu and then Brahma. The inactive part m signifies Yoga Nidra of Vishnu, which is a neutral state or inertial state. Goddess Kali supplies kinetic energy taken from the potential energy of Vishnu. Details of this opposing force or inertia are given later.

The model of our Universe in ancient times was developed as Big Pran (or life) or as Paramatma giving birth to Jiva Atman (child). This is like an unborn child resting in the womb of the mother and floating in the bed of water, i.e., Narayan. The unborn child is in a cavity or a large oscillator, which is the body of the mother and her womb. Heat is supplied from the mother's body to the unborn baby. Heat is life and comes from the mother as a

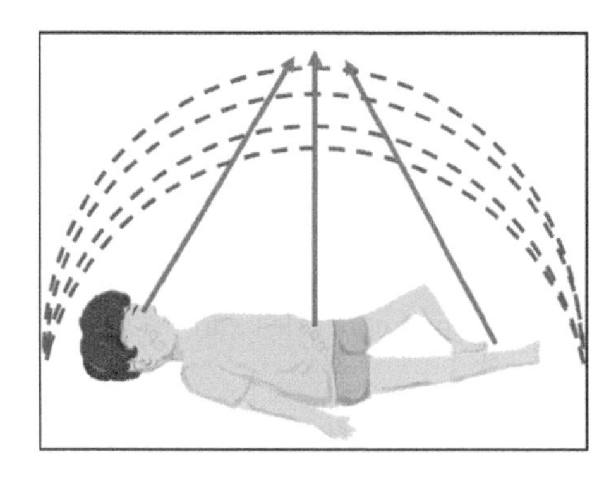

FIGURE 3.4
Creation of a focal point and vision by bending the space. Vishnu is watching the creation of curved space. The division and folding of space lead to singularity (Brahma).

fire to the baby. One oscillator creates another (first oscillation). Pran is the respiration process of the child. Before birth, the body rotates from a small spherical body to a full body with hands and legs, which is like the unwinding of a sphere into a flat surface. Air intakes give oscillations in the unborn baby. Finally, birth takes place with the first sound, i.e., Ma, Ma, Ma, a type of crying. This is Brahma who gets separated from the origin and carries the knowledge of the mother (and the father). Before birth, the body of the child develops two hands and two legs, which find four directions. In addition to the head, it gets five elements like Shiva. It is sleeping but dreaming in the womb. It is covered by the womb of the mother, which is Tamas in the darkness. Yet it has a life of fire. It can move within the space floating in the blood (Rajas) or red water. This movement including rotation is Rajas, the sleep or dreaming is Tamas, and the life of fire is Satwa. The body spreads out with time, it becomes bigger and bigger as Vishnu wanted and alone splits the body into hands, legs, head, etc. When the body started moving for the first time, it is Brahma. Coming out of the body from the mother is the rebirth of the second birth. So the love between the mother and father is transformed into a union/consort, which gives life (prana) to the parents' life. It follows duality since the mom or woman is undetermined at the first stage of birth followed by adopting three Gunas.

The stages of creation can be described as starting from a dormant state or vacuum having (almost) no but very massive vibrations. It is necessary to create the Sun or sunlight, which will bring life to Earth (e.g., sourcing water, etc.). So The Goddess must come out as the potential energy from the inactive body of Sri Vishnu from various parts that focus on a point and create a bright Sun. The massive flat body of Vishnu will rise and create a dome-like space that will accommodate the Sun, and from other forms of energy, Vishnu will create 'matter'. So this is a conversion of mass to energy to mass. This process is a competition to occupy space whether it is filled with vibrations or noise or disorder or a pure state. Overall, the heat should be spent effectively to raise the temperature, which should destroy all fluctuations that prevent the formation of plasma or condensate or the Sun and that bring order to the Universe. This needs the bending of space into a ball (sphere) or Brahmanda, as said in many places in Rigveda. The Goddess is a strong attractive force that can bend the space into a point like a black hole. In a condensate or the dormant state of Vishnu, it can be a vortex-like state (Tamas). In a flat space, it is a single flux that is split into n numbers, i.e., φ/n. It can rotate and move in the material freely through tunneling. The tunnel current is regarded as Brahma (Satwa), which is like the dream of Vishnu that can move everywhere even though it is static. By bending the space, another phase (φ) is created. This model is similar to the center of mass of the whole Universe, which is a collection of many zeroes. But for the sake of creation, the proper distribution of mass is necessary for creating several centers of mass. This is like creating many zeroes from one big zero (Mahasunya). These zeroes can be concentric plus distributed so that the space is occupied by the zero to maintain harmony. One god should be everywhere through the distribution of zeroes. The microcosmic or the center of the Universes is not manifested, but the macrocosmic Universe, including the centers of galaxies or solar systems (corresponding to smaller zeroes), is manifested (visible). Some have split themselves into different centers like $0/n = 0$, creating spots of light (fire) from a single, unique source of light. There is always a battle of duality to occupy the space between a (+) or a (−) part, matter or void, filled vs. unfilled space. To break the equality of a third object, a mediator is needed. Filled space can be considered as white (full of joy) and unfilled space as black (lack of joy), and there is an in-between space that is partially filled. These three states are Satwa, Tamas, and Rajas. However, no state is static, and they mutate to other forms S $\rightarrow$ T $\rightarrow$ R $\rightarrow$ S. This duality of white to black creates interference. The Sun is independent

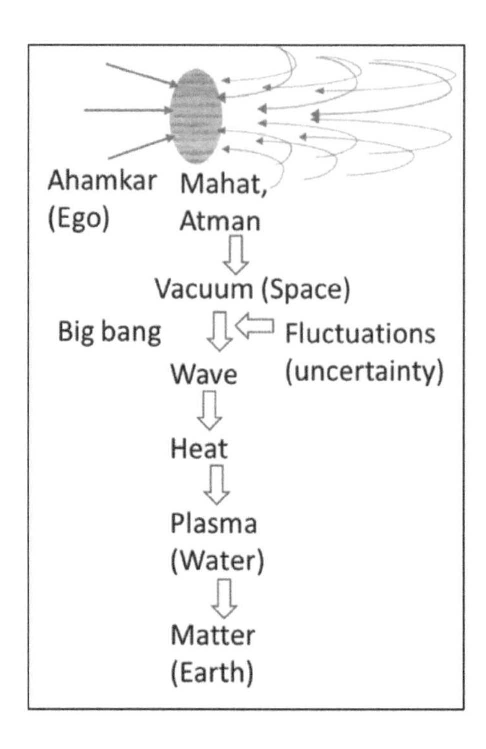

FIGURE 3.5
A flowchart of creation shows different stages starting from Ego and ending with Matter. An egg is created from three elements which give rise to many other elements. Various steps of the evolution process described in the Sankhya pilosophy are compared with nature and the water cycle. Creation starts from a ball of fire (or a vacuum), which is formed by combining three characteristics. Fluctuations of the vacuum or waves are created. This fluctuation can produce heat, which results in a blast (bang). Finally, the ball of plasma is dissociated into matter.

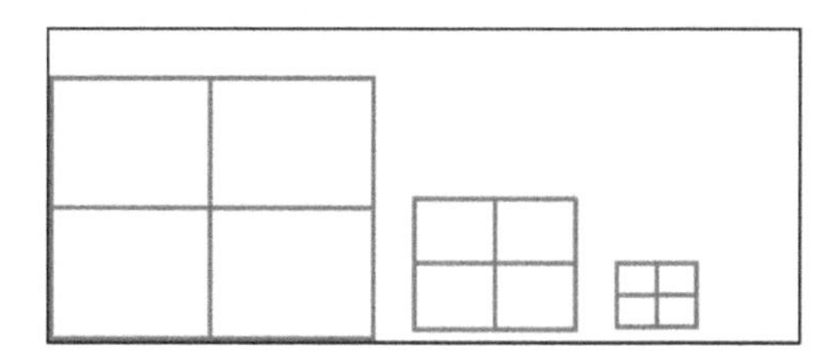

FIGURE 3.6
Folding space can increase the mass. The area of the square is decreasing with time and the density of the system is increasing.

of three states but surrounded by them, like the dawn, noon, sunset, and night, which is a cycle that occupies the space with three stages (like the three legs of Vishnu). To map a spherical space, three elements, e.g. γ, θ, φ, are needed or three axes, as in a gyroscope. This will keep the Sun as a static object, and either of the states can see the Sun all the time. This is also similar to three eyes watching the Sun continuously. It starts from red (Rajas) at dawn and dusk to white (Moon, Satwa) and dark (Tamas by night). So the transformation from black to white takes place through a state, i.e., red, i.e., love (Rajas).

Summary

In the previous chapters, we presented the creation of the elements such as Agni, and here we concentrate on the space or inflation of space. In this chapter, we presented a detailed mechanism for the creation of a large structure. Vishnu is the main god in this discussion, who produces Shakti, the primordial energy for our life. Other chapters will be closely connected to this very important chapter, which deals with creation. We have shown how to create a multiverse (or a metaverse) through the creation of zeros and a collection of zeros. In particular, a model of fluid is described in this chapter that will be addressed in ensuing chapters of the book.

The Universe is created from plasma or fluid mixed with positive, negative, and neutral particles. This is the condensed state of matter that is present everywhere as the cause of creation. It is like natural water, which gives birth to life on the planet Earth. Assuming water as the ground state, the model of creation in different cultures was put forward. From the beginning, water is considered the creator (God) who takes three forms, e.g., ice, water, and vapor, representing the three Gods Shiva, Vishnu, and Bramha, respectively. Both ice and water remain static, and vapor rises by the application of heat as described in Vishumaya. Heat plays with the three forms of water and the gods as a phase transformation. Heat (Agni) initiates life on the planet by creating a water cycle. In this case, water behaves like a man (Purusha), and Agni is woman-like (Prakriti). Both live within everyone and maintain life. Heat creates bubbles in the water from small to large sizes. They are hollow spheres or circles that resemble the egg model.

4

Theory of Evolution—(Sankhya) Philosophy of Dualism: Creation of a Cycle

In the previous chapter, I introduced some concepts of the creation of the universe, nature, conflicts between gods and demons, and gods as (apparent) creators. What happens if the creator becomes inactive and nature dominates over the creator? Such a model was conceived by seer Kapila many years ago, perhaps from his intense observation of nature.

Origin

In the beginning, the Earth was floating on the bed of water in the form of a limitless ocean. Then mountains were formed, but there were no rivers. Great sages from India were observing the mighty Himalayas, which are quiet and steady, full of ice like Shiva. Water vapor or clouds continuously hit (struck) him and were converted into rain through the condensation process. At some stage, the heat of the area changed, and a nonequilibrium process started. Some ice melted, and the holy river Ganga evolved from the top of the mountain. Ancient sages knew this from their ancestors, which inspired them to formulate the model of cycle, static associations with dynamics (Purusha/Prakriti), condensation, and life that comes from heat. Seer Kapila observed these changes very carefully, including the transformation of water from the ice that flows and then transforms into vapor, a perfect cycle of evolution. Having spent a long time in the Himalayas region, he followed the flow of the Ganga river and wanted to see the final form of Ganga when She merged with the ocean, which is also quiet (at a standstill). Seer Kapila moved to the Bay of Bengal, an island, and observed the tides created in the sea or ocean. The tides appeared to him like the heads of snakes, a multiple head system produced from the quiet ocean. Heat created water vapor from the ocean and then clouds. They were transmitted to the mountains, the Himalayas. In this process, rain or river Ganga was created and maintained. This process is described as 'Vishnumaya' (discussed in the previous chapter, 3). Love changed the world of interactions as well as philosophy. Learned people from the desert areas did not have the opportunity to see snow-capped mountains to the end of long rivers or black clouds in the sky that bring rain to the Earth. They developed the idea of a flat Earth floating on the water, only a static universe. Because of love, Prakriti (a woman-like Goddess or nature) repeatedly interacted with Purusha (a man-like supreme god). She went back to

DOI: 10.1201/9781003304814-5

FIGURE 4.1
The statue of Seer Kapila is worshipped in the temple, the delta of the River Ganges, West Bengal, India.

FIGURE 4.2
A dualism model explains interactions. (left) The Union of God and Goddess. (right) Mother Mary and Jesus Christ.

Purusha and released her energy, did not merge with Purusha permanently, but rather came back to Purusha through periodic cycles since She possessed some repulsive force. She kept her separate force.

Attractive and repulsive forces coexist like love and hate and complement each other. If a particle wants to move by the influence of an attractive force, then it will always face a reactive force that attempts to stop the free creation, and then it is a back reaction or a repulsive force. If love is a force, then it works between two or more so it starts from a dual nature. This energy is like heat, which excites particles from the ground state. When this energy is spread out, then the particle becomes de-excited, tries to return to the ground state, and again becomes excited. This work mimics the Sankhya description of Purusha–Prakriti interactions in the upcoming text and in the diagrams.

Basic Principles

This chapter deals with the mystery of creation, which is described as the spread of knowledge or consciousness. Mass–energy equivalence has been described as the basic model of God. The concept of energy in the void of space is proposed, which interacts with an absolute static object, the Purusha, as the only source of knowledge or consciousness. Although India was great in mathematics and the ancient scriptures were written based on some astronomical observation (and measurements), very little mathematical treatment has been given to support the rules or laws of nature. 'Sankhya' (or 'Samkhya') in Sanskrit means number. In many places, numbers are used as in the seer Kapila's Sankhya philosophy, where the equations seem to be hidden. It is distinguished by its emphasis on the enumeration of creative principles. The two principles of Purusha and Prakriti are its foundation. This is followed by the principle of three Gunas. The five organs of perception, together with manas (mind) and ahamkara (ego), make a total of seven. Manas, together with organs of perception and organs of action, make a total of 11. These 11, together with ahamkara and buddhi, make a total of 13. Ahamkara and manas, the ten organs of perception and action, together with the five subtle elements of sound, touch, form, taste, and smell, give a total of 17 principles. These, together with Buddhi and Prakriti, give us 19. Finally, the number of principles evolving out of Prakriti is 23.

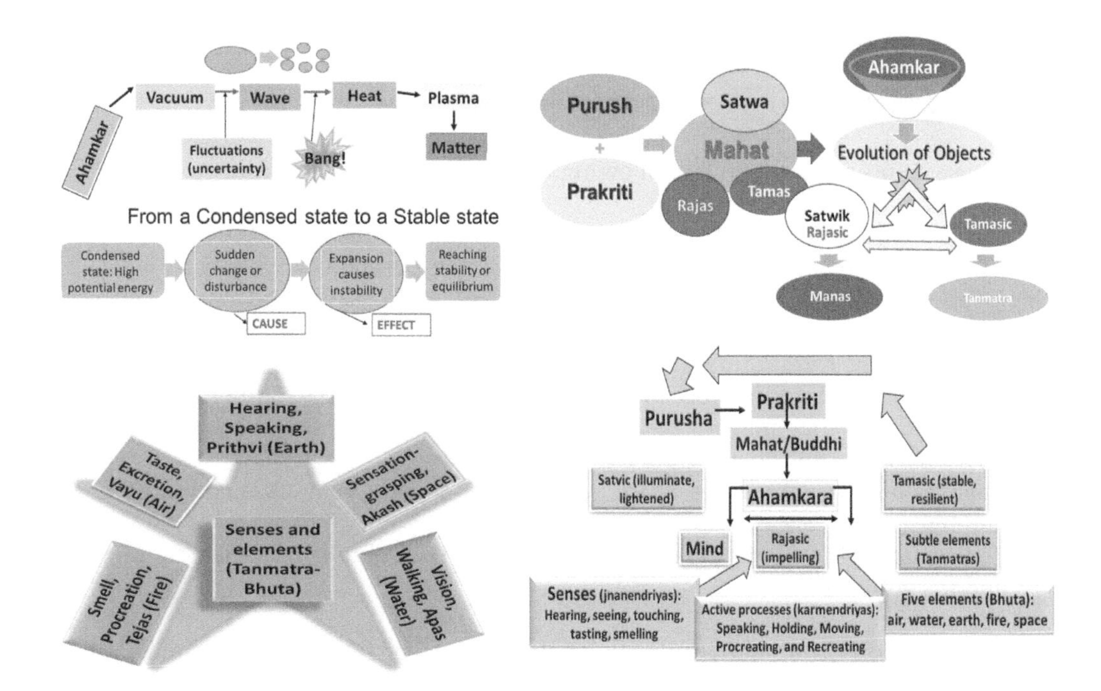

FIGURE 4.3
The evolution of the world according to the Sankhya philosophy shows a network of entanglement. A flow chart shows different steps of creation. One is split into two objects and so on. There are 24 steps in total. The last step shows splitting into five levels.

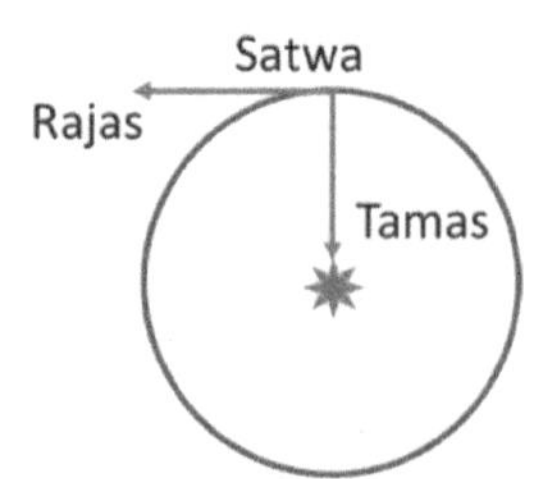

FIGURE 4.4
A cycle is created as a combination of three opposite characters: Satwa, Rajas, and Tamas. A circle shows two directions of velocity; one is tangential and the other one points towards the center.

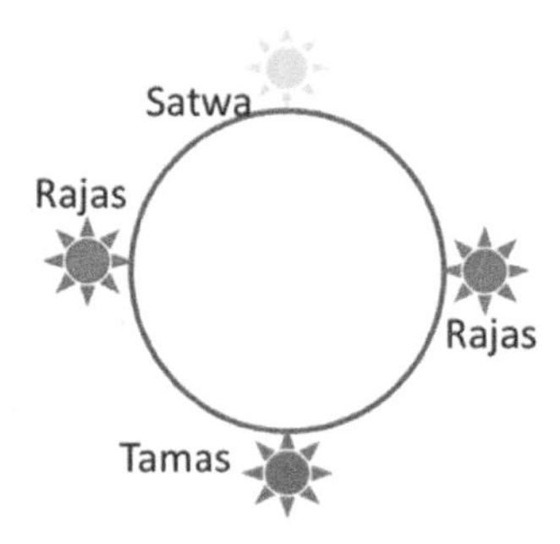

FIGURE 4.5
The cycle corresponds to the Sun, who changes colors from the sunrise to the sunset. A circle surrounded by four suns from different directions. From the left, the red color sun becomes yellow at the top of the circle. On the right side of the circle, the sun appears to be brown which becomes blue at the bottom of the circle.

The main points are given in the following:

- Purusha has no attributes except that 'it is' and that 'it knows'.
- "The spirit is what it sees, it is isolated, indifferent, a mere inactive spectator."
- It has no intelligence (this is located in Buddhi), and it is without desire. It is pure freedom.
- Originally, the three Gunas (Sattwa, Rajas, and Tamas) are in perfect equilibrium in Prakriti. But under Purusha's influence, nonequilibrium and evolution begin.
- Sattwaguna is the 'stuff' of consciousness and all higher mental states (associated with the Hindu god Vishnu and the goddess Laxmi).
- Rajasguna is the source of activity, sensation, and emotion (associated with Brahma and the goddess Sarasvati).
- Tamasguna is the source of resistance, inertia, and dissolution (associated with Shiva and the goddess Kali).

The static Purusha is the source of knowledge (or heat) that always radiates light. Prakriti collects energy from Purusha by creating circles (voids or Sunya) around Him. Circles of different sizes are collected in a very small volume as Mahat, which creates Buddhi. At the critical energy or size, this gives birth to more circles in a very symmetric manner. The addition of three Gunas produces circles of different shapes and sizes. These are called 'atoms' consisting of several orbits in a small body that is invisible. When many atoms meet, we can see them physically as fire, water, or matter or experience the effect like air or the sky or free spaces (voids). Sankhya philosophy is not derived from *Veda*, which considered Sun to be the only creator. Sankhya considered a creator, particularly the Sun, not a creator but an inactive source of knowledge. Seer Kapila wanted to look at the universe, which included the Sun, Earth, Moon, the five planets (Mercury, Venus, Mars, Jupiter, and Saturn), and even other stars as atman. Five visible planets work like five organs, senses, or elements. This metaphysical idea still holds today.

Seer Kapila gave birth to the science of the mind, which is the foundation of all philosophy. It explains Vishnumaya (or great illusion) as a strong attractive force that is essential for the creation of any kind of object, including the curved space and expansion of the universe. This force separates cause and effect as the universe is completely active and inactive. This active part is dynamic or natural, which cannot be manifested but maintained in a state of equilibrium. This maintains different levels of energy, the lowest one being Tamas, which works as a deep potential or attractive one. The second one is repulsive, corresponds to a higher state of energy, and is called 'Rajas'. The highest state of energy is Sattwa, which maintains an equilibrium of strong attractive and repulsive forces. However, the state of equilibrium can be destabilized, and again it starts to regain the equilibrium. This process creates the emission of waves as light. Ultimately, Brahmanda or waves are emitted in a waveform that has a periodic form: a cyclic form of creation and destruction with an intermediate state of stability. This reminds me of the creation of stars from a cloud of particles, and the death of a star is associated with the emission of radiation.

Out of the great book, *Sankhya Karika* by Iswarkrishna, I selected 20 shlokas from 67 shlokas to be discussed here. Most of the text here I have taken from this book; however, I have made some additional explanations based on possible applications today. The problems discussed are fundamental and still applicable today.

Like many other models, Sankhya's philosophy is also based on the solar system being developed in ancient times. The people of Earth observed that the Sun and Moon were reappearing every day and that they appeared to be revolving around our planet Earth, assuming that the Earth is a static object. So we have two skies or spaces created by the Sun and Moon in the geocentric model. However, if we fix the position of the Sun, we get three orbits for the Moon and Earth, including rotation and revolution. These three types of orbits are eternal, as stated in the first shloka. This is the first heliocentric model in the history of science.

Analysis of Selected Shlokas

In Shloka 1, dissatisfaction (dualism) is described in three types. We have two states, which create dualism or confusion and which are the main cause of dissatisfaction. Three circles can be formed for complete satisfaction.

Shloka 2 directly offers the definitions of Purusha and Prakriti as unmanifested and manifested. The Sun, being so bright, cannot be visualized, whereas the Moon's face can be seen easily. The difference between Purusha and Prakriti or the separation between them can cause incompleteness of the circles, as stated in the second shloka.

In Shloka 3, Purusha is clearly defined as being beyond cause and effect, which lie at the center of the circle. The fundamental Prakriti is a perfect circle without any defect or discontinuity around the Purusha, as stated in Shloka 3. Also, seven slightly defective circles and 16 heavily defective circles surround the main Prakriti and the Purusha, which complete the 25 elements of the Sankhya philosophy. In the third shloka, number seven can be

FIGURE 4.6
The creator with many hands and several heads, sitting on a lotus bed, represents the Sun. A statue of Buddha on the lotus seat. He has six main hands and several smaller hands which point in different directions. Few smaller heads can be seen on top of his head.

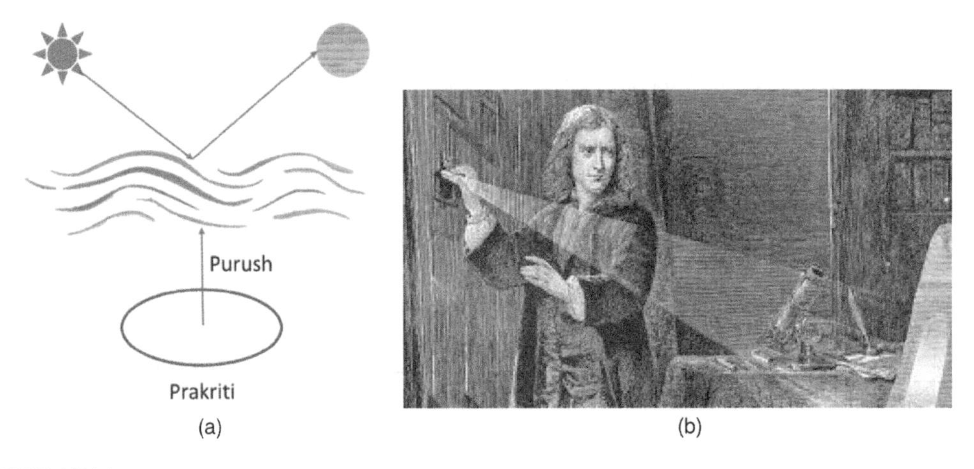

FIGURE 4.7 (a)
The sunlight interacts with the waves on the water and forms an interference pattern, called illusion or Maya. It is compared with Purusha and Prakriti.

FIGURE 4.7 (b)
(right) Sir Isaac Newton using a prism to spread the light. He demonstrated how to break light into its constituent parts by refraction and how refractive analysis creates phenomena like the rainbow.

associated with the span of a week consisting of seven days or a seven-color spectrum of the Sun, i.e., VIBGYOR. Sixteen can be associated with the phase of the Moon. On day 16, a full Moon becomes a new Moon.

One of the revolutionary comments is found in Sankhya as the definition of evidence or experimental proof. Evidence can be found in three types: (1) Direct, which is observed, (2) guess or hypothesis, and (3) what is said. These three types remind me of three lokas or three forms of Agni (or water). The first one is our Sun, which is directly observed; however, all the activities of the Sun cannot be seen. The second one is the Moon, which reflects light from the Sun and creates enormous effects on Earth. Since these two are not associated with the sound, we need other evidence, i.e., lightning or a thunderstorm, to demonstrate the evidence of the effect of the Sun.

Shloka 4 can be interpreted as the effect of light–matter interactions that produce sound. The second form of evidence is Anuman (guess, perception), which is based on the finding of some objects or Anu. This raises the concept of reception, or Indriya. This evidence is also of three types, namely start, end, and intermediate, which needs a detailed study. How do we find fire in a place where there is no apparent source of the fire. One has to search for it through indirect evidence using all our senses like smell, taste, touch, in addition to the eyes and ears, i.e., vision and sound. So we need five organs: eyes and ears for the direct observation of fire and three indirect ones, nose, tongue, skin (body), to feel the heat or fire.

We know Atman is unmanifested like heat, but how can we see him? We have to taste him through our senses if he is invisible or inaudible. Our brain always applies a signal or intelligence to identify a new object and appoints all kinds of senses that can capture low to high frequencies from light (electromagnetic) to touch (mechanical), i.e., from terahertz to subhertz frequencies. This is like understanding cooked food; starting from the change of color, to a boiling sound, to the temperature of the food, and finally to the taste of the food. Before dawn, there is no light; we wake up from the sound of the birds. The Sun sets with the sound of birds. The heat of the Sun burns our skin. We feel the heat affecting our body temperature. Buds bloom in the morning and produce smells. The smell is created by heat. All senses are connected to the brain. The signal is captured in a vacuum created somewhere in the brain. Filling up the space gives us pleasure like a thirsty person drinking water. When we smoke or inhale some nice perfume, we feel joyful.

In the Sankhya philosophy, there is no term called 'insufficient' or 'Abhaba'. The opposite of joy is fear, or Bhoy. Fear has three types: night, dark cloud, and eclipse, all of which produce darkness. Light also has three types—Agni in the sky (the Sun), lightning, and fire on Earth—which gives us happiness or courage. To produce fear or Bhoy, we need the absence of light, which is the main cause of unhappiness (Dukha). It is always there. There is no need for the removal of darkness as is said in the first shloka of the *Sankhya Karika*.

Shlokas 5 and 6 describe types of guesses. Is it a direct or indirect process? Can we analyze facts through direct observations or based on the senses as indirect information? Prakriti and Purusha are not observable since they exist in a subtle body; however, they exist. We can feel the existence of Prakriti through work or kinetic energy. More subtle objects mean that an ultrathin object is acting as a tunnel barrier, as stated in Shlokas 7 and 8.

Shloka 9 makes a strong statement about the definition of cause and effect. Unlike other schools of philosophy, Sankhya philosophy says that both cause and effect are Sat, or real truth. Before the effect starts, it is hidden as the cause in a subtle body. Kinetic energy is an integral part of potential energy, which is manifested as work or dynamics.

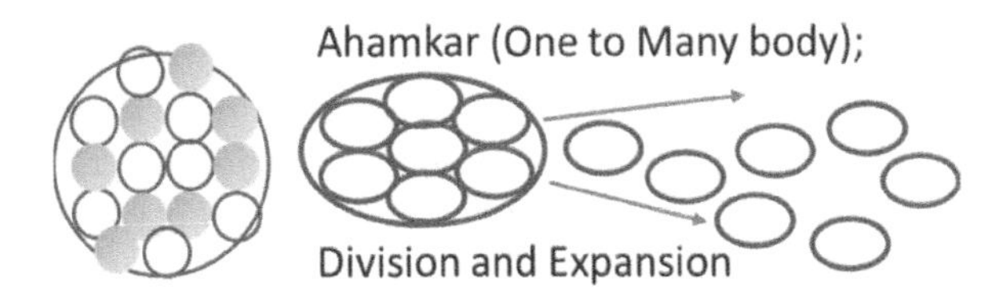

FIGURE 4.8
A zero and many zeros describe the universe, including the atomic orbits, which can be extended to vortex structures. Many zeros are created from one big, absolute zero. Dividing one large void (zero) n times, n number of zeros can be created. This can be described as a big circle containing several closely packed small circles. After the explosion of the big circle, the small circles go outward directions.

FIGURE 4.9
A giant wheel called the `Sudarshan chakra' is placed on top of the Jagannath temple, in Puri, India. It has eight spokes. The rim of the wheel is decorated with eight symbols that look like a flame of fire.

Imagine that a massless light particle is emitted from the Sun and is trying to reach a solid matter, i.e., Earth. It has to penetrate different layers that have different transmission coefficients. The first one is the space between the Sun and the Earth (sky or vacuum) (see Figure 3.5). Then air, vapor (ionosphere), and water layers are to be overcome to reach solid Earth. A massless particle gains mass through interactions in these layers. Mass is gained through overcoming resistance in these layers. The *Mula Prakriti* is extremely thin. Water is less subtle since it can be pushed. The layer of Tejas, or ions, is like a dense gas that can be seen but cannot be touched properly even though it affects our bodies. Air is more subtle since it cannot be seen; however, it touches our bodies. More subtle is the sky, which cannot be touched or pushed; however, it produces sound or waves. Ultimately the cause of all effects is the most subtle object, which is massless. A massless particle cannot perform any mechanical work. To do so, it has to gain mass or become *Sthula'*. Alternatively, a massive object becomes massless through the emission of energy. So massive and massless particles can transform into each other. Whether visible or not cause and effect exist, they transform into each other. This is what is meant by creation in Sankhya philosophy, i.e., transformation explained in Shloka 9.

Shloka 4 says all manifested matter has a cause and is time-dependent, i.e., not free from the life cycle. However, Mula Prakriti is not manifested. She is free from any cause, unlike the manifested one, and is completely disconnected from Mahat. She depends on herself and creates her work. A seed cannot be transformed into a plant if it is not assisted by water, air, soil, heat, and space. However, the work of Mula Prakriti is completely different and produces Ahamkara. She is like the Sun or fire who produces her heat. She is the heat in the body that transforms the form. She is the only driving force: consciousness. She lives in the body of Vishnu and is emitted from the body like tunnel currents that can penetrate any barrier. She comes out of the body spontaneously without depending on any potential difference. Most likely through the breaking of a body, this energy is emitted like a nuclear reaction or the beta decay process.

Ancient people knew that the Sun, the Moon, planets, various stars, constellations, and even galaxies (Milky Way) consist of molecular dust. From routine observations of the heavenly bodies, the spherical shape of the universe and orbits in the solar system has been understood. The world is described as a shell or multiple shells, which is a large void filled with nothing. Objects like the Sun and Earth interact with each other, influencing the idea of dualism. The two objects might be the Sun and the Moon (or a combination of the Earth and Moon). Later, the trinity connects Sun-Moon-Earth as a part of creation-preserving destruction. Although this model around the Sun as the object was popular in ancient civilizations (Sumer and Egypt) with the Sun at the center and accompanied by two other entities on either side, Sankhya for the first time unified and extended the idea of the trinity. He brought the concept of energy into the void of space that is interacting between the Sun and Earth. The Sun is our father, and energy is our mother or nature or Agni (the fire or the flame). Ma is not only mother but also Mon (mind), Maya (attraction), which variably describe our mother, generally not available with the father. This concept is essential for modeling the creation of our universe to include the mind.

Let's look at nature and understand 'creation' in the light of the Sankhya philosophy. At this state, since all energy is spent, the sky tries to cool down and form itself back to the solid state containing closely packed atoms. The size of the spheres would be the smallest. Also, visible solid matter becomes transparent (as liquid), then air with a state of radiant plasma, and then completely invisible in the form of the sky, which is filled with energy or soundwaves without any solid matter. The formation of rain and stars in the sky can be explained. In the crude form of energy, we say that the eye, ear, nose, tongue, and skin are all representations of circles that collect sensations of different frequencies. Five forms are distinct light, sound, smell, taste (heat), and finally touch that intercept frequencies from the high frequencies of light to merely low mechanical frequencies. Touching may produce friction and increase the heat energy. With the increase in the rate of touch or friction, heat energy increases, and finally fire (light) can be produced. After the fire is ignited, heat is spent, and things can be cooled down to the frictional force, which is at best invisible, opposite to fire which is most visible. Now eye, ear, nose, tongue, and skin can also represent circles or closed spaces that confine different frequencies.

Nature is described as the dynamics or the mechanism of movements of a body, as representations of atoms or voids. Collecting different frequencies from light, sound, smell, and heat stops with direct contact or the lowest frequency, which is touch. Touch produces interactions of the lowest frequencies; however, by increasing frequencies, higher energy states can be produced. These five frequencies or forces represent interaction that collects energy or information, which is like potential energy. Kinetic energy or activities can help to store energies as positive energies if they form circles or a close space through movement. To grab an object, we need hands. To rotate a body, we need two hands. While we run or walk, our legs form cycles. To express something through voice, we need to obey our mouth, which creates space. Some of the limbs or organs in our body are hidden, and energy is produced. In a crude sense, all four organs in our body produce activities of different kinds, but they are linked to one where all activities come to an end. These activities store heat in our body, which is released from our survival. All are manifestations of periodic contraction and expansion, digits to small circles. They are all connected to the nonequilibrium state of Gunas, explained as binding and unbinding like a snake that makes vortex-like structures from a point to an uncoiled snake and goes back into an egg.

We have described Purusha as a circle with a strong attractive force. However, Purusha can be many, like many circles that can have an infinitely large number. The world is a reflection of Purusha. Purusha is extended everywhere (c^2), with no birth or death. It has no end or limit concept that was developed. It is like a catalyst that ignites the reaction by increasing the energy but does not actively constitute the reaction. It has no intelligence and is free from perception. Purusha is always bigger (much bigger) than Prakriti. Prakriti is described as a collection of energies or small masses (particles). They try to become a large entity collectively having extracted energy from the source of a large energy, i.e., Purusha. Prakriti tries to compete with Purusha and ultimately merge (or transform them into) Purusha. Purusha loses the attractive force and develops some weak repulsive forces,

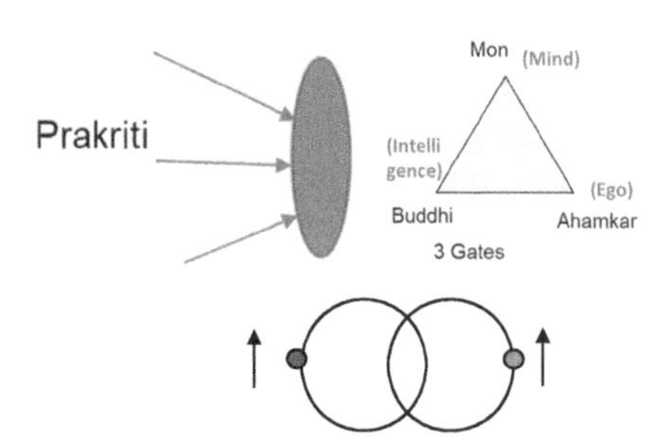

FIGURE 4.10
(Left) Nature (Prakriti) is a combination of three different characteristics. Three arrows are converging to a lens-like object. Three sides of a triangle represent three different characters. Two circles are rotating clockwise and counterclockwise intersect. (Right) The holy trinity as observed in a church. Light appears through the cloud.

which are termed His wish to split. The lack of force of Purusha is transformed to Prakriti, whose state of equilibrium changes, and it splits into three energy levels since Purusha and Prakriti are in the form of two circles (a doublet).

Purusha is originally a big circle, but a small circle derived from it (Maya) can also be described as Prakriti. This doublet creates a triplet state. However, these energy levels try to equalize in order to attain a state of equilibrium or a collective state of energy: a superposition of states, an entangled state that creates lots of energy in a condensed form. This state is created through fluctuation, i.e., attraction–repulsion that emits energy or force at different levels—primarily four forces (energy) like gravitation, electromagnetism, strong, and weak forces, which are very distinct in strength. However, they finally react to an ultimate force that is like the limitless sky and acts as a source of all four forces. This is like Lord Shiva who has five faces. The fifth one is a reflection like Mother Kali or Ganga, which is shown on the head of Shiva. It is also shown as a snake on the head of Lord Vishnu.

It is understood that Seer Kapila discovered the mechanism of rotation, i.e., moment and angular momentum as a combination of Purusha and Prakriti, as a dimension. The center of man is Purusha. The relation proven can be explained by centripetal/centrifugal forces with a central point (static). So these energy levels can be formed or can further split into 4+1 energy levels. A shell model of the universe can be described as consisting of 25 levels. Jumping from one to another level costs radiation (the emission of light). Purusha at the center contains all potential energy like gravitation, which is strongly attractive. Prakriti transforms it into other forms. This is an atomic shell model developed one century earlier. Five means 4+1, four directions on a plane, and one normal to the plane. Two Sattwa Gunas consist of five knowledge and five activities (work). So (4+1)*2 = 10. Eight forms, eight directions on the plane, and two up and down. Understanding the fifth element and division of nature in the unit of five can be very challenging, which may be avoided in the following discussion. We have just touched the elementary level of the five-base divisions here and focus on the dualism in the following section.

Views on God

1. If the existence of Karma is assumed, the proposition of God as a moral governor of the universe is unnecessary.

2. Even if Karma is denied, God still cannot be the enforcer of consequences because the motives of an enforcer God would be either egoistic or altruistic.
3. Despite arguments to the contrary, if God is still assumed to contain unfulfilled desires, this would cause him to suffer pain and other similar human experiences. Such a worldly god would be no better than Samkhya's notion of a higher self.
4. Furthermore, there is no proof of the existence of God. He is not the object of perception, there exists no general proposition that can prove him by inference, and the testimony of the Vedas speak of Prakriti as the origin of the world, not God.

Therefore, Samkhya maintained that the various cosmological, ontological, and teleological arguments could not prove God.

Connection with Other Philosophical Thoughts: Plato: Sankhya (Trinity)

Nature of the Soul

To obtain knowledge, one must suppress bodily needs and concentrate on rational pursuits. The job of the rational component is to postpone and inhibit immediate gratification when it is for the best long-term benefit of the person.

The nature of the soul is as follows:

- Comprised of three parts (tripartite)
- Rational component (Satwa)
  - Immortal, existed with the forms
- Courageous (emotional or spirited) component (Rajas)
  - Mortal emotions such as fear, rage, and love
- Appetite component (Tamas)
  - Mortal needs such as hunger, thirst, and sexual behavior that must be satisfied

Plato: World of Ideas

- Pure, ideal form (order)
- World of senses
- World is an imperfect copy of Mind.
- Illusion: Shadow of Ideal form (like Brahma and Maya)
- Sun as Fire
- Color forms by sunbeam splitting
- Sun (a form of God)
- Intelligence (eye)
- Knowledge (sight)
- Forms (visibility)

Aristotle

- Life as a hierarchy in which God is omnipresent
- Infinite god (like infinite souls in Sankhya)
- Cause and effect

- Theory of causation (Sankhya type)
- God: Full form (Purusha)
- Manifested–nonmanifested
- Sufficient (full)–insufficient (empty)
- Excess–deficiency
- Nondualism (like Vedanta philosophy)

Aristotle on Motion in a Void (Zero)

- Believed impossible:
  (i) no medium to sustain the motion
  (ii) absence of resistance would lead to infinite speed, an unacceptable solution
- Classification: Two different parts of the world:
  - There is the world all around, where things come and go; are born, live, and die; and motions start and stop.
  - There is the world up in the sky, where things happen over and over again: the Sun rises and sets, the seasons reoccur, the planets repeat cycles.
- Sankhya combines Plato and Aristotle.
- From flat (two-dimension) space to curved space
- Division of zero, singular to multiple

Thales: Water

- Prime element
- transforms vapor—rain
- Creates waves
- Earth on a water bed
- Interference—dualism

Anaximander: Boundless

The boundless form of materials and no form of material.

Anaximenes: Air as Element

- Movable
- Changeable

Heraclitus: Fire

- Formless fire like Atman (unchanged)
- Movable and changeable like the river
- Produces illusion like a wave (Maya)
- Attraction (love)–repulsion (hate) and Nous (mind/intellect)

Pythagoras: Number

- Number is God.
- $C^2 = A^2 + B^2$ can be explained from trinity and laws of equilibrium.

Socrates: Sankhya + Naya philosophy

- Sun: Form of God lies beyond being
- First form of Mystic Theology
- God-Evil Duality; Uncertainty
- Not both are known at a time even for Gods.
- Like Sankhya: Free from a creator

Decartes: Dualism

- Illusion (like Brahma and Maya)
- Meditation
- Consciousness

Spinoza

- Body and Mind
- God is total
- God is cause of everything
- (like Vishnu in the ocean of Cause)

Kant

- Space and time dualism + Cause (from experience)

Hegel

- Thesis, Anti-thesis and Synthesis
- (particle, anti-particle and Light)

Schopenhauer

- Universal will = Nirvana (Satwa) free from desire (Rajas) and Suffering (Tamas)

Neitzsche

- Against Christian concepts!
- God is the sum total of three things

Leibnitz

- Truth of two types: Reasoning and Fact
- Universe is our mind
- Monads
- Space and time duality

Newton

- Equations of Motion
- Law of Gravitation

Faraday-Maxwell

- Duality of Electric and magnetic fields
- Electromagnetic waves

Quantum Mechanics

- Schrodinger: Entangled states
- Heisenberg: Uncertainty principle
- Dirac: Particle-antiparticle

Einstein

- Mass energy dualism
- Holographic universe

Christianity: Christian philosophy emerged around the 3rd century to reconcile science and faith, starting from natural rational explanations with the help of Christian revelation. Many philosophers believed that there was a harmonious relationship between science and faith though there is a contradiction and others tried to differentiate them. The development of Christian ideas represents a break with the philosophy of the Greeks, bearing in mind that the starting point of the Christian philosophy is the message of the Christian religion. The ideologies in this sector make religious convictions, rationally evident through natural reasons – the attitudes are determined by faith in matters relating to cosmology and everyday life. But the co-existence of Christian philosophy is questioned as this philosophy itself depends on revelation and established dogmas. However, Christians accept the holy trinity (God as the Father, the Holy Spirit, and the Son of God or Jesus Christ). Famous painters depicted many characters around God and a dualistic character (Mother and the Child). The Christian symbols are decorated with the Sun at the center of the holy cross and the moon. The soul does not suffer from death which leaves the mortal body at a high altitude, however, returns to Earth in due time and activates a new body.

Buddhism: Like a dormant God in Sankhya Philosophy there is no god in Buddhism, yet Buddhists offer prayers to several gods and goddesses as Hindus do. The creation was not in Buddhism as everything emerged from 'nothing'. All states are considered equal which is like the superposition of states. Lord Buddha searched for the origin of 'unhappiness or grief' as we find in Sankhya Philosophy in three different forms. He followed a 'middle path' which is the co-existence of two states. The 'eight paths' of life seem to be related to the dissolution of dualism by the application of the 'trinity'. Buddhism is found to be very close to the 'Upanishads'. The "Karma" is the active energy of consciousness. Karma can be described as a combination of 'Mind', 'Body', and 'Sound' like Rajas, Tamas, and Satwa, respectively as described in Sankhya Philosophy. Re-birth is one of the main pillars of Buddhism which signifies a periodicity of nature. Nothing is permanent or constant in the universe which moves in a cyclic motion. Like goddess Kali, Goddess Tara brings a transformation of the world. As things are created, they are dissociated regularly. The net result is a zero.

Fire, dualism, modern science: Ancient people after creating fire tried to preserve it. To store it they split the fire and keep the burning objects separately. One form of fire is calm and quiet (slow burning) which can be vigorous while using (burning) a body and an excited state (fast burning). The fire was considered as a collection of small bright particles which are synchronized and can move like a wave when interacting with the air. During the movement of the very excited fire flames are split and the flames compete, like a bunch of serpents. Small particles come out of the excited fire. Scientists tried to understand the nature of fire from a steady state to an excited state.

I believe that Isaac **Newton** was the first person to suggest the quantum theory of light consisting of light particles as opposed to waves. Indeed, the nature of the particles was not known at that time. As the developer of differential and integral calculus, Newton played with the differential equations of waves. However, he gave a revolutionary idea of the particle nature of light which is the foundation of quantum theory that we use today. Although Newton's research on fire has not been published, I believe he used different sources of fire (light) for his experiments on 'light'. Using a prism, he split (white) sunlight into seven different colors. Very similar works have been practiced in ancient times with fire. In particular, the fire was described as a combination of seven flames, and each flame was assigned to a particular form of goddess. The seven colors of light can be assigned to seven orbits that emit light at different wavelengths. They can be described as seven particles of different energies. Each level or orbit has a different magnetic phase which can be split by an external magnetic field as experimented by Michael Faraday. Each orbit has a different area and associated solid angle or a geometric phase. Agni or fire is a ball of plasma that consists of charged (ionized) particles or dipoles which can be affected by a magnetic field. Faraday split flames with a magnet in different orientations since each level has a particular magnetic phase.

If Newton would have introduced light particles, it is Michael Faraday who added 'polarization' to the light particles of Newton. Depending on the polarization of the particles i.e., positive and negative ones a loop or cycle can be assigned to the particles either clockwise or counterclockwise. They can be mutually perpendicular and hence complementary to each other. If one of them represents an electric field the other one is a magnetic field since both are rotating cyclically. One of them can be fixed and the other rotates around the fixed center. This model matches the dualistic description of nature as described in Sankhya's Philosophy as the interactions between the motionless observer (Purusha) and the rotating Prakriti. They represent the co-existence of a ground and an excited state.

Albert Einstein studied the excited state and developed the photoelectric effect. In the theory of relativity, the rest and the moving frames can be linked to two cycles that are closely linked. They constantly observe each other, hence a relative motion is developed. This can be extended to the spin-orbit coupling as a combination of two cycles aligned perpendicular to each other. The dualistic nature of fire can be described as containing (or supplying) heat and being manifested as light; however, these two forms are inseparable.

Selected Shlokas

1st Shloka: "DuhkhatrayabhigatatJijnasa Tadabhigatake hetau/ Drste sdpartha cetnaikantatyantatabhavat".

Meaning: Sadness is of three types, with its persecution the wish emerges for knowing the way for cessation of grief. If it is said that the way of seeing in a scriptural way is meaningless, I would say no, it is not meaningless because, from the viewpoint of a visible manner, there is no way to the cessation of grief, and grief cannot be eradicated forever.

2nd Shloka: "Drstebadanusrobikoh S Hyahbisuddhikkhyatisoyojuktoh/ Twadwiparitoh Shreyan Byaktabyaktojnabijnyanyat"

Meaning: The works enjoined or prescribed by the Vedas are compared to the way of seeing because there is impure mortality and disparity. Therefore, which is opposite to it, that is impure immortality and indiscrimination, i.e., wide beyond the range in that manner. The stated–nonstated knowledge of men is the reason for that manner.

3rd Shloka: "Mulprokritirbikritirrmmhyadyadyadh Prokritibikritoyoh Sapta/ Shoroshokostu Bikaro N Prokritino Bikritih Purushah"

Meaning: The original nature is not the distortion of anyone; the seven materials of the world, like greatness etc., is the nature distortion that is also nature itself, as well as the distortion; 16 materials are purely distortional, and the soul of men is neither something natural nor the distortion.

4th Shloka: "Drishtamanumanmaptabachanang cha Sarbaparamansiddhatat/ Tribidhong Promanmishtong Promeyosiddhing Promaniddhi".

Translation: Proof is of three types—ocular (direct), assumption, revealed truth (words) —because every proof is included in these three forms. From the proof, the admission of evidence is determined.

5th Shloka: "Protibishoyadhyabosayo Drishtong Tribidhomonumanmakkhyatam/ Tollingolingipurbokomaptosrutiraptobachanantu".

Translation: From worldly affairs and sensual pleasures, assiduity (intellect) is created; that is the direct proof. The intellect from the extreme suit and religious knowledge; that is the assumption. It is said that assumption is of three types and that, for the revealing statement, knowledge from the content of a sentence is proof of words.

6th Shloka: "Samannyotosto Drsihtadotindrinang protitinumanang/ Tanmadapichasiddhong Porokkhomaptagmat Siddhom".

Translation: The elements that are transcendental, foresaid, slightly ocular are validated by assumption at the end. The indirect elements are not validated by assumption but can be validated by revealed words.

7th Shloka: "Otidurat samipyadindriyo-ghatanmonohonbosthanat/ Soukkhmodbyabodhann=advivobat Somanaviharacchho".

Translation: By (1) far distance, (2) far nearness, (3) lack of sensibility, (4) unmindfulness, (5) subtle, (6) demarcation, (7) censure, (8) mixture with comparable substances, and (9) feelings; incomprehensibility of the elements is created.

8th Shloka: "Soukkhmyat Todonupolobdhinavashat Karjyatostuduplobdhe/ Mohodadi Taccha Karjyong Prokritiswarupaong Biruponcho".

Translation: The male nature is not direct; by this, the subtlety of those elements is created, and any myth is not related. With the help of work, the existence of nature can be felt by assumption (the feeling of the existence of male is determined by a different type of assumption)—that action is of the greatest nature. All of those actions are similar in nature and in some cases dissimilar too.

9th Shloka: "Asadkaranadupadanogrohonat Sarbasambhabavabat/Saktassya Sakyakaranavabacchha satkaryam".

Translation: Work is honest. There is the existence of work before the source because one that is honest—that has no existence—can never be produced or generated. Second, if each and every work can never be generated, then it is said that, with work, the cause has a definite alliance (no relation can be generated with work having no existence; to maintain alliance work is to be considered honest). With the definite effect of energy, for certain works, the energy-alliance combination is the effect of works. (In works having no existence, the alliance of energy is impossible; thus even if the alliance with work is accepted, then with the request of energy alliance, work is to be considered honest.) Third, work is not different from cause (thus can work be considered honest?).

10th shloka: "Hetumodonityombyapi Sokriyomonekomasritong Lingam/Sabayobong Porotontrong Byaktong Biporitombyaktom".

Meaning: The intention of Sankhyacharyas is that the quality of goodness or excellence is venial and revealing, that appearance regarding the qualities of rajas or menstrual excretion is functional, and that pioneer and worldly delusion is the spiritual guide or preceptor and covering. This triplet is the sensual pleasure for men as well as the way of salvation and, by fortune, just like a lantern with fire-wick-oil combined, accomplishes self-oriented works.

11th Shloka: "Sattvamlaghu prakasakamistamupastambhakam chalancha rajah/ Guru Varanakameva tamah, Pradipvaccaarthato vrittih".

Shloka 11 starts describing three Gunas. All manifested and nonmanifested objects are different from Mula Prakriti. However, Purusha or Atman is different from both manifested and nonmanifested. Liberation comes when the separation between Purusha and Prakriti becomes a unification. It is like two opposite things that are annihilated and produce energy. Again we use the model of the Sun and Moon. The Moon is lifeless or unconscious. Sometimes it is completely dark or nonmanifested. The meeting of the Sun and Moon means eclipse.

12th Shloka: "Prityapritivi§ddatmakahPrakdsapravrttiniyamarthah/ Anyonyabhibhavasraya–Jananamithunavrttayasca gunah".

Satwa gun is related to happiness, Raja stands for sadness, and Tama is related to delusion. For Satwa gun, displaying manifestation is required, for Raja predilection is required, for Tama engrossing is necessary.The propensity of these three Gunas is that they overwhelm one another, are sheltered by one another, are the cause of origin and also friendly with one another.

13th Shloka: "Sattwang Laghu Prakashakmishtamuposhtombhokong Rajah/Guru borornkomeb Tamah Prodeepboccharthoto britti".

Meaning: The contemplation is that Satwa is trifling and the publisher, Raja is active and instigating, and Tama is the guru being covered with protection. These tri-Gunas are a pleasure for the male and reason for Mukti (liberation). By virtue, just like fire-accomplishment-lantern with oil combined, they execute their own work.

17th Shloka: "Sanghatpararthatvat Trigunadiviparyayadadhisthanat/Purusoqosti bhoktribhavat Kaivalyartham pravrttes-ca H".

Meaning: The existence of men is to be considered because it is the conflict substance that is not the thing for conflict but something similar to it. It is the requirement for others; men are something excess to the conflicting substances. The supreme One is the consumer and instinct for salvation. The existence of men is to be considered even for these three reasons.

20th Shloka: "Tosmat twatsongjogadchetonong Chetanadib lingam/Gun-kotritte cha totha korrteb bhabatyudasinong".

Meaning: The unconscious or inanimate object like greatness is adjacent to the consciousness of men, though the power of authorities is not with men who are fully adjacent to the men as an authority.

21th Shloka: "Purusasya darsanartham Kaivalyartham tatha pradhanasya/ Pangandhabadubhoyoropi Songjogstotkritang Swargah".

Meaning: The ideology related to masculinity is a majority for the spiritual heart. The blasphemy is related to divinity.

23rd Shloka: "Adhyabasaya Buddhirdhormmo Jnyanaong Birag Oisharyam/ Swatwikmetadrupong Twamsomsmadwiporjyostom".

Meaning: Perseverance or assiduity is the intelligence—that is, the support of intelligence. Religion, knowledge, abstinence, and riches (pomp) are the saintly characteristics of intelligence. Its opposite characteristics are iniquity, lack of knowledge, nonabstinence, and lack of pomp and are all misappropriate to intelligence.

24th Shloka: "Abhimanohohongkarostasmat Dwibidih Prabartate Swaragah/ Ekadoshkascha Gonostanmatraponchokonchoibo".

Meaning: Arrogance or huff is vanity. From this vanity, the dual work, that is, the eleventh sense and five subtle elements or rudiments, is evolved.

36th Shloka: "Eteprodipakalpah Porosporbilokkshona Gunabishesha/Kritsnong Purusharrthang Prokasshya Buddhou Projchanti".

Meaning: External, Mind, and Vanity are the elements that go with the three Gunas. Though they are not interrelated with one another, they are masculine like the lantern; expressing the whole thing by handing over to growth.

67th Shloka: "Samyag Jnyanadhigomadhormmadwinamkaronpraptou/ Tishtati Sang skarbosachchkrovromiboddhritasoriroh".

Meaning: With the theoretical conversation, religious diversity can't be a reason for sensual pleasure. In that case, even if Kulal-matter does not exist, in a culture of high speed, with the movement of the Kulal-disc, religious diversity exists for a certain time in the body. This completes the understanding of a Wheel or a Chakra.

Summary

Sankhya philosophy is traditionally viewed as an atheistic philosophy as it accepts the authority of *Vedas*. Sankhya is strongly dualistic and has been theistic or nontheistic, with some late atheistic authors, such as the author of Sankhya Sutras. The word 'Samkhya' Sankhya philosophy is one of the six astika schools of Hindu philosophy. This philosophical idea was developed by Prajapati Rishi, Angiras Rishi, Gautam Rishi, and six sages who composed *Rigvedas*. Seer Kapila was influenced by Gautam (founder of Naya philosophy). This philosophy is addressed in Chapter 2 (among others) of the *Bhagabhat Gita*. Gautam Buddha studied Sankhya philosophy. It is not purely metaphysical, but a logical account based on the principle of Conservation, Transformation, and Dissipation of Energy. The Dualist system later was criticized and essentially dropped by Tantric and Vedanta (nondualism).

Science, religions, and cultural traditions develop theories and creative descriptions of the origin of the universe and the meaning of life. These theories have both similarities and differences, regarding the cause and effect of creation as well as for life as human beings know it. Religions and cultural traditions primarily adhere to a personal God as creator and ruler. Science has gone in the opposite direction of denying the existence of a God. A definitive cause of creation has not been scientifically found. Science may find a comparable, suitable match in the ancient thought of Sankhya, written in 500–800 BCE era. Sankhya is probably the first complete philosophical description of the origin and evolu- tion of creation. The three basic energetics of Sankhya are comparable to the basic ener- gies of physics (light, kinetic, and mass). This addresses the hypothesis that the evolution and origin of creation stem from the three energetic Gunas (light, motion, and inertia) that make up what is referred to as materiality or nature Prakriti, being described in Sankhya thought.

5

Interactions at the Macroscale—Classical Mechanics—Equilibrium and Rotation: One Cycle

We discuss the following topics in this chapter:

- **Newton's laws of motion:** Force, inertia, dynamic equilibrium
- **The motion of the planets:** The law of gravitation
- **Special relativity:** Mass–energy equivalence (dream of Vishnu)
- **Angular motion and Vortex:** A model of the Universe (Brahmanda)

In Chapters 1 and 2, we presented various models of the creation and the creator(s). In Chapter 3, we described a complete process to reach equilibrium (a stable state) through the circular motion of elements from a liquid-like substance, which can also show the creation of space. In Chapter 4, we discussed static equilibrium and the Sankhya philosophy. The creation of mass can be found in the mythological story through the interactions between gods and demons, as described in Chapter 3. There is another story of creation where Shiva, his consort Sati, and Vishnu interact with one another (see Figure 8.7b in Chapter 8). Shiva became angry when Sati died, and he hurled Her body (space or Tamas) and rotated (creating a phase or curved space) with his property of Rajas. Finally, Vishnu had to divide the space (after rotation) into several pieces. In this way, this process leads to stability. It is believed that the shape of the Universe is like a flame of a fire that goes to a sharp point like the flat part of the symbolic model of Shiva (called a 'Shivalinga'). The linga part of Shiva is Brahma, which is an oval seat on the flat part (Yoni) of the bed of Vishnu, which is space-like and can be broken by Shiva/Goddess Kali within the body. Initially, the flat part is spread out arbitrarily due to the excess repulsive force. Due to the duality, the goddess applies strong attractive forces to bring all scattered particles against the repulsive force created from the dualism. If there is no mass, there is no energy gap. To create mass, glue to bind particles is needed, This creates an energy gap. One of the best ways to create strong connections is to condense matter and break space or convert to curved space. This is described as Brahmanda, the holographic Universe. Brahman is watching its creation. This oval or spherical shape of the Universe can concentrate sound waves and create resonance like a lens or a blowpipe where sound circulates in the bent space. Pure sound comes out as Brahma, as resonance, and as Goddess Kali and circulates air in the constricted space. Resonance stops the noise. Resonance is the signature of holography. This process is described in the chapter 'Purushottam Yoga' in *Sri Bhagavad Gita* as a tree with roots in the sky.

In this chapter, we explain mechanical motion and a dynamic equilibrium based on the discussion given. Here we describe the motion of dynamics of a particle, which can

DOI: 10.1201/9781003304814-6

be treated by the circular motion. The curvilinear motion can be described by Newtonian mechanics, which explains planetary motion. We describe dynamic equilibrium based on three forces overcoming the duality or the interplay of the mass (inertia) through the change of speed of an object with time. Sankhya's philosophy also gives a general understanding of Newtonian mechanics and a three-body interaction. Sankhya's philosophy describes creation as a combination of three different characteristics, which are also attributed to three forms of deities (God or Goddess). I try to relate this to the foundation of classical mechanics since both routes relate to the equilibrium and dynamics of massive particles.

In this chapter, we shall learn the force of action manifested by the change of velocity with time (or acceleration) of a massive object. It has a particular direction, and many of the forces can be added to find the resultant or net force. When it applies to the body, it tries to resist as a reaction force. The object has inertia or mass. If the body does not move, then the action and the reaction are equal and opposite, and the system remains at equilibrium. So the net applied force is zero. How do we quantify force? It is the capacity to move objects and causes physical change. It is resistance like friction, or it overcomes the resistance. It works against natural tendencies like slow-to-fast motion or vice versa. If the force causes an action, the change of momentum represents the effect. Manifestation of interactions is given by forces that cause the time rate of change of momentum.

The force changes the state that can be felt with time. It works like heat that causes a change in velocity (or temperature). The temperature change can be compared with the change in the momentum of an object. Momentum can be described as the net applied force multiplied by the time the force is being applied. The action and reaction work in pairs between two bodies, hot and cold. The force of impact is governed by the speed and inertia of the object. The interaction time is crucial for the measurement of momentum. The longer the impact (interaction) time interval, the smaller the force of impact. Momentum can be measured from the change of resistance or inertia and displacement. Momentum is the most important quantity in mechanics (instead of energy), which consists of a static quantity (space-like), and a dynamic quantity, which is a measure of change in space. Mass is the cause of space bending or occupying space. Force accounts for both space and change in space and time. In a free space that is open and undisturbed, force cannot be applied. Inertia or mass occupies the space and creates curvature. As a result, a barrier is created that resists the mass's motion. Hence, force is applied to overcome the resistance. As the mass starts moving, there is a π-phase change in the action. We learn the basic concept of classical mechanics from some of the greatest scientists, Galileo, Newton, Kepler, and others. As in India, ancient Greek philosophers developed the concept of a force. Similarities between the ancient Sankhya philosophy and Greek philosophy have been observed particularly in the usage of duality and trinity; however, the concept of force was clearly stated in the Sankhya philosophy. The main flaw in the concept of force in Greek philosophy was corrected by Galileo and later by Newton. The ancient Greek philosopher Aristotle had the view that all objects have a natural place in the Universe, that heavy objects such as rocks wanted to be at rest on the Earth, and that light objects like smoke wanted to be at rest in the sky and the stars wanted to remain in the heaven. He thought that a body was in its natural state when it was at rest and that, for the body to move in a straight line at a constant speed, an external agent was needed continually to propel it; otherwise, it would stop moving. It is Galileo Galilei who realized that a force is necessary to change the velocity of a body, i.e., acceleration but no force is needed to maintain its velocity. The tendency of objects to resist changes in motion was what Johannes

Kepler had called 'inertia'. This insight was refined by Sir Isaac Newton, who made it into his first law of motion, also known as the law of inertia: No force means no acceleration, and hence the body will maintain its velocity. As Newton's first law is the restatement of the law of inertia that Galileo had already described, Newton appropriately gave credit to Galileo. As we see, one of the three Gunas, Tamas, represents inertia in Sankhya philosophy. Of course, two other Gunas, Satwa and Rajas, correspond to a constant (steady) motion and acceleration, respectively. Three Gunas are the main constituent of Newton's laws of motion. Let us understand these three laws of Sir Isaac Newton.

In classical mechanics, Newton's three laws of motion describe the relationship between the motion of an object and the forces acting on it. Mechanics is the study of how objects move or do not move when forces act upon them—a lesson proving that immovable objects and unstoppable forces are the same. The three laws of motion were first compiled by Isaac Newton in his *Philosophiae Naturalis Principia Mathematica* (*Mathematical Principles of Natural Philosophy*), first published in 1687. Newton's first law of motion states that an object either remains at rest or continues to move at a constant velocity unless it's acted upon by an external force. Newton's second law states that the rate of change of momentum of an object is directly proportional to the force applied or that, for an object with constant mass, the net force of an object is equal to the mass of that object multiplied by the acceleration. Newton's first and second laws of motion are valid only in an inertial frame of reference. For objects and systems with constant mass, the second law can be restated in terms of the acceleration of an object. Newton's second law can be written as the net force applied $(\boldsymbol{F}) = \mathrm{d}\boldsymbol{p}/\mathrm{d}t = \mathrm{d}(\mathrm{m}\boldsymbol{v})/\mathrm{d}t = \mathrm{m}(\mathrm{d}\boldsymbol{v}/\mathrm{d}t) = \mathrm{m}\boldsymbol{a} = \mathrm{md}^2\boldsymbol{s}/\mathrm{d}t^2$. Here, $\boldsymbol{s}$, m, $\boldsymbol{p}$, $\boldsymbol{a}$, and t represent the displacement, the mass, the momentum, the acceleration of the body, and the time, respectively. Thus the net force applied to a body produces a proportional acceleration. This great law combines three opposing elements: The force or Satwa (like knowledge or wisdom) is a product of mass (inertia or Tamas) and the acceleration (Rajas). It also states how one can overcome inertia. If the mass signifies the distribution of space, the acceleration corresponds to the time double derivative of the space of displacement. If the force is attractive, the mass acts as repulsion, which opposes the attraction or vice-versa.

Newton's third law of motion states that every action has an equal and opposite reaction. The third law states that all forces between two objects exist in equal magnitudes and opposite directions. If one object (A) exerts a force ($\boldsymbol{F}_\mathrm{A}$) on another object (B), then B simultaneously exerts a force ($\boldsymbol{F}_\mathrm{B}$) on A; in this case, the two forces will be equal in magnitude and opposite direction. Therefore $\boldsymbol{F}_\mathrm{A} = -\boldsymbol{F}_\mathrm{B}$. The third law means that all forces are interactions between different bodies or different regions within one body. Thus there is no such thing as a force that is not accompanied by an equal and opposite force. In some situations, the magnitude and direction of the forces are determined entirely by one of the two objects. The force exerted by an object (A) on an object (B) is called 'action', and the returned force exerted by an object (B) on another object (A) is called 'reaction'. The third law of Newton is sometimes referred to as the action–reaction law with $\boldsymbol{F}_\mathrm{A}$ as an action and $\boldsymbol{F}_\mathrm{B}$ as a reaction. In other situations, the magnitudes and directions of the forces are jointly determined by both bodies, and it is not necessary to identify one force as an action and the other force as a reaction. Both forces are part of a single interaction, and neither force exists without the other. The two forces in Newton's third law are of the same type (e.g., if the road exerts a forward frictional force on the tire of an accelerating car, then it is also a frictional force that Newton's third law predicts for the tires pushing backward on the road).

The Motion of the Planets

The force of gravity causes the moving planets to travel in roughly circular orbits around the Sun. They have been circling the Sun for billions of years because other forces have been too weak to change the orbits in any significant way. Just as the planets orbit around the Sun, they rotate around their axis, a line running through the center of the planet; thus it can be said that the planets follow rotational rules of motion. All planets including the Earth move in two ways. Earth moves around the Sun, completing one solar orbit in roughly 365 days. The planet also spins on its axis, rotating completely every 24 hours. Other planets take different times to do the same things, but they all go through the same types of motions. While it is common to think of the planets of solar systems as moving in circles, the orbits are ellipses—oval—rather than perfectly round. If they were circles, the Sun would be at the exact center of the orbit. That's not true for ellipses, though for most of the planets, it's pretty close. The orbit is the result of two competing forces, the straight-line motion of the planet through space pushing against the gravity of the Sun. Planetary orbits and rotation were the results of the creation of the solar system. The Sun and planets were born to form the collapse of an interstellar gas cloud. When the vast cloud shrank and solidified, it did so with enough force that it put the bodies of the solar system in motion.

Newton was convinced that the planets must obey the same physical laws that are observed on Earth. This meant there must be an unseen force acting on them. He knew from an experiment that, in the absence of an applied force, a moving body will continue in a straight line forever. The planets, on the other hand, were moving in elliptical orbits. He realized that the answer was gravity—the same force, that causes an apple to fall to the ground on Earth. In the time of Sir Isaac Newton, everything that was known about planetary motion could be summarized succinctly in three laws attributed to Johannes Kepler. The first law states that the planets move around the Sun in an elliptical orbit. The second law states that a planet sweeps out equal areas at equal times. According to the third law, the square of the orbital period is proportional to the cube of the distance of the Sun. These are purely empirical laws, which describe what happens without explaining why it happens. Newton developed a mathematical formulation of gravity that explained both the motion of a falling apple and that of planets. He showed that the gravitational force between any two objects is proportional to the product of their masses and inversely proportional to the square of the distance between them. It can be expressed as $F = G m_1 m_2 / r^2$ with the product of two masses (m_1, m_2) and the interatomic distance (r). When applied to the motion of a planet around the Sun, this theory explained all three of Kepler's empirically derived laws.

Let me repeat the Visnumaya that was described in Chapter 3.

> I, Brahma was sleeping and did not feel any attractive or repulsive force. I did not have any sense of love, i.e., I was cold. Heat and love were induced in me as if I started attracting parts of my body. With this attractive force, I reveal a repulsive force as its dual as Prakriti. But I tried to change Prakriti into me through interactions.

So repulsion becomes attraction and went back to repulsion again. This cycle continues. When the repulsive force becomes much greater than the attractive force, then the creation process begins. Dissolution starts when an attractive force becomes much stronger than a repulsive force. Equilibrium is established when two forces are equal. This is the origin of Tamas (attraction), Rajas (repulsion), and Sattva (equilibrium). Attraction or repulsion

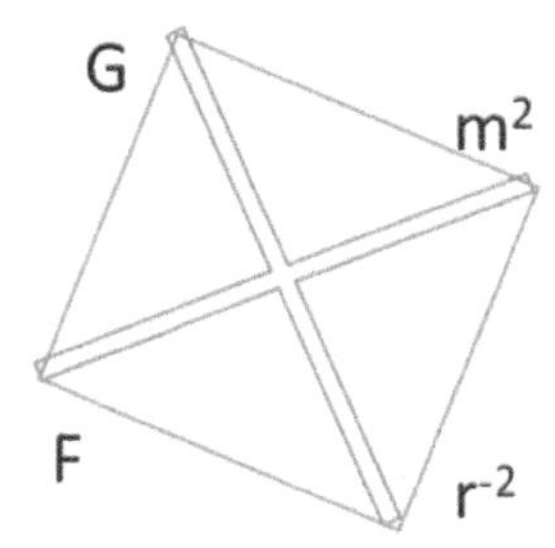

FIGURE 5.1
A quadrature shows the relationship between forces, mass, distances, and the gravitational constant at four corners. Four apexes of a square represent four different characters used in Newton's law of gravitation.

works with an opposite force, which works like friction and produces heat and fire. Love is a kind of interaction that produces heat. It works like fire or Agni. (This principle should apply to electric and magnetic fields that work perpendicularly to each other and will be discussed in Chapters 6 and 7). Gravitational force is also combined with centrifugal and centripetal forces. If we rotate the Earth at a high speed, then things will move away from the surface of the Earth, gravitational attraction can be overcome, and Jiva can be liberated (Mukti).

At that time, there were many gods as creators of the world, and they fought each other claiming their superiority. God was believed to be like a man with supreme power. There was also a concept of the Sunya zero, or nonexistence; however, that philosophy was not dealing with energy or a process rather than a static world, and a concept of action and reaction was lacking. So Seer Kapila disregarded the concept of an active god as a creator, as well as that of the zero filled with energy or elements. He argued that Purusha is completely inactive. It is like a ray of light that is not affected by any external force or field. Newton formulated his first law of motion as a particle moving in a straight line in the absence of other external forces or force fields, i.e., Prakriti or nature, which is completely decoupled from Purusha. So Purusha is an (absolutely) static object and has no reaction or deformation. Then how can we measure force? Seer Kapila introduced the idea of dualism. The action of Prakriti has an equal but opposite reaction. So Prakriti or Purusha-Prakriti manifests as dualism. This is regarded as Newton's third law of mechanics. Now, it is most important to experience the reaction that gives us knowledge of nature when we apply an action that is nothing but a change of velocity with time. This action of Prakriti is different from Purusha since he is inactive, but he can still oppose the action by applying a resistance, i.e., inertia. Prakriti must overcome the coupling to Purusha, i.e., inertia to start an action. So reaction to the impulse is a product of the coupling constant (i.e., inertia, m) and the action (rate of change of velocity). This is the famous second law of motion in mechanics proposed by Newton: $\boldsymbol{F} = m\boldsymbol{a}$, a process consisting of only three elements. Action and reaction can be like two zeroes that are coupled with inertia, resistance, or back force, which is a constant. So from a dual nature, we get three elements that exchange with one another. This is a measure of force, and now we would like to measure it without any action or independent of time.

Let's say Purusha has a great mass; it cannot move but can attract or repel other masses. The attraction depends on the distance between Purusha and Prakriti, which has a mass of say m_2. As they come closer, the attraction increases, which is the inverse of the distance between them. Now reactions and actions are dual, and they affect them equally. So

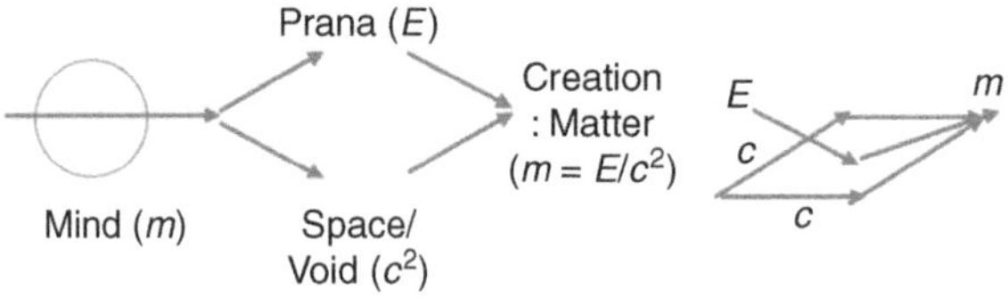

FIGURE 5.2
A connection between energy, space, and mass. Mass is split into energy and space. Mind is split into two parts that rejoin into a point and the creation starts. The energy strikes a plane and produces mass.

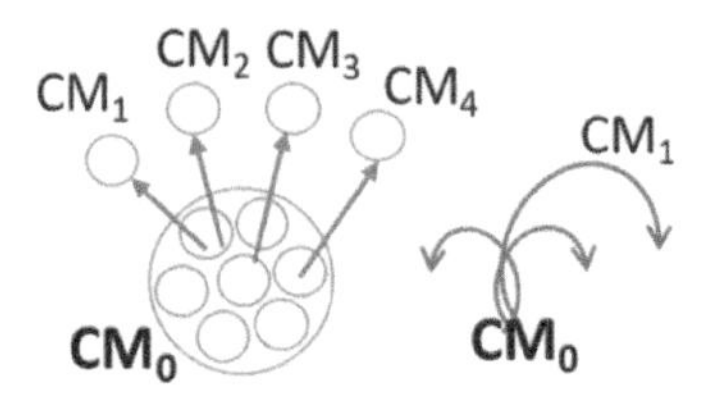

FIGURE 5.3
Center of mass is defined as a combination of distribution of masses; however, it is not always manifested. Moving the CM is the key to the development of an efficient machine. It gives a projection of the body to another place. A circle containing closely packed circles explodes and the small circles go out from the center and create different centers of mass.

the force is a product of m_1/r and m_2/r. Again, assuming Purusha is inactive and resists this action, a coupling constant can be imposed on the process. In Newton's famous law of gravitation where Purusha and Prakriti are considered equally (in fact all particles are equally treated), this constant is a Sattwa that creates a balance between action and reaction, i.e., G (m/r_1) (m/r_2) or $F = G\frac{m_1 m_2}{r^2}$, where $G = 6.674\times10^{-11}\ m^3 \cdot kg^{-1} \cdot s^{-2}$. (The same logic applies to Coulomb's law of repulsion, $F = \frac{kQ_1Q_2}{r^2}$ and will be discussed in Chapter 6.)

So far, we understand that the model of the Universe is very complex and consists of many geometries and patterns that go beyond Newton's laws of motion. Newton's three laws of motion hold to a good approximation for macroscopic objects under everyday conditions. However, Newton's laws of motion, combined with universal gravitation and classical electrodynamics, are inappropriate for use in certain circumstances, most notably at very small scales, at very high speeds, or in very strong gravitational fields where space is curved. Therefore, the laws cannot be used to explain phenomena such as the conduction of electricity in a semiconductor, optical properties of substances, and superconductivity. Explanation of these phenomena requires more sophisticated physical theories, including general relativity and quantum field theory. In special relativity, the second law holds in the original form $\boldsymbol{F} = d\boldsymbol{p}/dt$, where $\boldsymbol{F}$ and $\boldsymbol{p}$ are four vectors. Special relativity reduces to Newtonian mechanics when the speeds involved are much less than the speed of light (c).

Special relativity is a theory of the structure of space–time. It was introduced in Einstein's 1905 paper "On the Electrodynamics of Moving Bodies". Special relativity is based on two postulates, which were contradictory in classical mechanics: (1) The laws of physics are the same for all observers in any inertial frame of reference relative to one another (principle of relativity). (2) The speed of light in a vacuum is the same for all observers, regardless of their relative motion or the motion of the light source. The theory has many surprising and counterintuitive consequences. Some of these are:

1. **relativity of simultaneity:** Two events, simultaneous for one observer, may not be simultaneous for another observer if the observers are in relative motion.
2. **Time dilation:** Moving clocks are measured as ticking more slowly than an observer's stationary clock.
3. **Length contraction:** Objects are measured to be shortened in the direction that they are moving concerning the observer.
4. **Maximum speed is finite:** No physical object, message, or field line can travel faster than the speed of light in a vacuum. The effect of gravity can only travel through space at the speed of light (c), not faster or instantaneously.
5. **Mass–energy equivalence:** In $E= mc^2$, energy (E) and mass (m) are equivalent and transmutable.
6. **Relativistic mass:** This is a simple idea used by some researchers. Thus the overall defining feature of special relativity is the replacement of the Galilean transformations of classical mechanics with the Lorentz transformations (according to Maxwell's equations on electromagnetism).

We shall explain these great laws/theories in the light of Sankhya's philosophy.

The basic elements are duality and breaking the duality, which can create mass and energy. Of the three elements, one remains constant, and the other two interact with each other. We have seen the application of Maya (Shakti) for the rise of Vishnu and the creation of Brahma with the help of the goddess Kali. Vishnu is inactive, like still water or condensed gas, with little vibration. Radiation excites Him to produce water vapor i.e., division of space by the application of heat (Kali) which goes up into the air and forms a cloud that covers sunlight (Madhu-Kaitava) and is finally destroyed by Vishnu to produce rain. Sun appears in the sky (Brahma). The demons Madhu and Kaitava represent a condensed state of the particles. Goddess Kali creates the division to produce mass from massless objects, which also creates light. This is the model of Earth. This can be applied to create

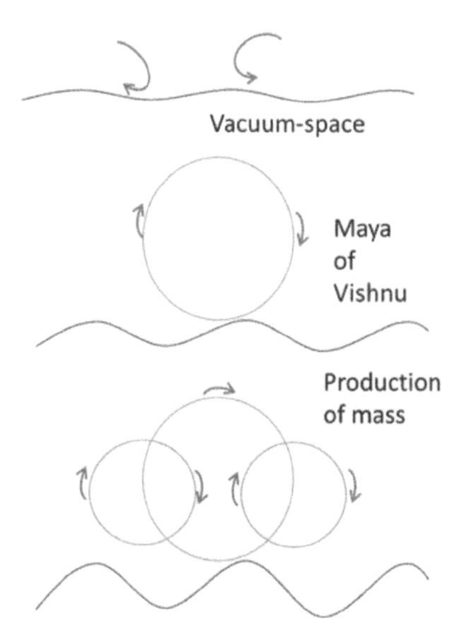

FIGURE 5.4
An interference pattern is created from maxima and minima of intensity or a duality of bright and darkness. Initially, there was an undisturbed line. A circle interacts with a wave-like line and produces two other circles on either side of the original circle rotating in opposite directions.

the Sun from gas in space by replacing water with the ionized gases. But the creation of the Universe with many objects like planets, stars, etc. requires a universal model, which is Mother Kali who wears a garland of heads (flowers from garlands). Goddess (Mother) Kali, being the supreme god, holds all celestial objects around her. She is the source of gravitational force. She kills massless objects and converts them into massive objects (like transforming gaseous objects into solid matter or vapor into raindrops). Vishnu does this for Madhu–Kaitava, which are doublets, e.g., attractive and repulsive forces or objects. From the complete objects, a neutral or different object can be formed. Ultimately, a nondual state is formed from the dual states. So we have three elements: Satwa (neutral), Rajas (motion), and Tamas (massive). Mass (m) and charge of motion (dv/dt) work oppositely, and when they interact (compete), the impulse or momentum (p) of a response rises. So the force $\mathbf{F} = m\,dv/dt = m\mathbf{a}$. Similarly, electric, and magnetic fields (E) complete one another and produce light. Electric is Rajas, and the magnetic field is Tamas because they are repulsive and attractive. The goddess is like a magnetic field or gravitational field. In the law of gravitation, m^2 works against r^2 (space), yielding the force $\mathbf{F}$. In the $E = mc^2$ formula, energy works against space (c^2) to yield mass (m). c^2 stands for the expansion of space, the attraction of negative space. Overall, Satwa-Rajas-Tamas interact with one another. They can be classified as follows.

Our mind, which covers Atman, strikes knowledge of fire constantly, produces life or vibration, and finally emits a sound (big/deep sounds like a blast). Our mind works like the Sun's rays, which excite water (or space) and create small to large vibrations. Water contains potential energy and heat (as fire lives in the woods as fossil fuel). In general, this is the story of the creation of the Sun. Inside the Sun we have two forces: One is nuclear, i.e., strongly attractive (like the Yukawa potential), and another is a weak repulsive Coulomb potential. Due to the repulsive force, matter cannot be created from the ionized gases. This strong attractive force overcomes the repulsion and matter forms. The matter is created from the interaction of energy (E) or vibration or heat with the sky (space, or c^2) because more matter is needed in less space. The addition of energy creates more space, which resembles the splitting of molecules by heat. This explains the mass–energy equivalence formula $E = mc^2$. In this formula ($E = mcc^2$), since mass is moving at the speed of light, i.e., time does not change. Here mass and energy are dual, and the space is a

FIGURE 5.5
Creation of a cycle, a Universe and also two other cycles, which gives the stability of the system. Three circles e.g., unfilled, semi-filled, and filled. The filled and unfilled circles are superimposed with an unfilled and filled quadrupole (fan-like) structure, respectively.

constant as c^2, which resists the conversion as it is a vacuum. Creation (m) = E/vacuum to $E = mc^2$. Ultimately, we divide a zero into two or three or into many nonzeroes, which react to one another and create a great zero at equilibrium. Splitting zeroes in higher numbers will solve complex processes that are unknown today! It is like bubbles created from soap water. In our dream, we produce a multiverse. An example can be found in the ancient texts (in *Atharva Veda* and *Vishnumaya*, as described in Chapter 3) and is explained again in the following text. Here, we shall interpret mind–matter interactions.

In our previous discussions, we see how Newtonian classical mechanics has been influenced by Sankhya philosophy, particularly in stating these laws of mechanics where inertia or mass becomes active in the presence of energy, and the acceleration of creation starts. Similarly, in electrical and magnetic forces, particles are charged or magnetic if electric or magnetic fields are applied. Most importantly, the mass–energy (or space–time) concept proposed by Einstein is very similar to Sankhya's. Consider that one of the most interesting parts of the Sankhya philosophy is the collection of mind or great mind or intelligence, which is the first manifestation of Prakriti, which is transformed into ego or soul and which is dissociated into five work-related senses and five knowledge-related senses and one mind and, most importantly, five Tanmatras, which is the smallest form of particles described as 'light', liquid, smell, touch, and sound, which manifest as Earth, water, radiation, air, and the sky. Everything is created from the sky (Akash), which is the main element called 'Matter', which is converted into energy (Sakti). Energy (E) and matter (m), or Sakti Akash, live together and form the collected or condensed state. They can interchange through space (c^2). One is visible, and the other is invisible, and, to transform into invisible, it should move as fast as light or the space created by the speed of light c^2, which also gives the dimension of energy.

So mass is converted into energy when it moves at the speed of light, and then energy changes to mass when trapped by an infinite space c^2. $E/c^2 = m$ means squeezing invisible energy into an infinitely large space or an infinitely small time (t). This is a state of great crunch, which is defined as the end of time or a period. It is the state of darkness when Purusha was inactive or in a dormant state. At that point, fire energy (E) strikes repeatedly on the space (c^2) and is condensed to the space. Oscillations at a slow rate to a fast rate produce different frequencies, and different materials are produced. Different frequencies mean different sizes of circles or orbits and different levels of confinement. For a continuous change, it can look like a vortex that starts from a large circle, goes to the smallest one, and then unwinds into a large one, like from defocusing to focus and back to defocus. However, the Sankhya philosophy describes this in five distinct steps that transform and show five entirely different objects, apparently as their energy is at different states. Starting from solid matter, liquid and gas and plasma states are formed as you increase heat, and ultimately space or sky is produced, the largest sphere you can imagine, which extends to infinity.

Dream of Vishnu

It is like creating a multiverse where each zero represents a Universe. In this way, we create images in our minds. Seeing myself in the mirror gives me enormous pleasure (selfie, photo shot). I feel I am the beautiful one. It gives me pride and develops my ego (Ahamkara). That is the origin of excitation and creation through the division of oneself. Atman is static, but the mind draws an image on it, and it propagates, assisted by intellect, and finally establishes our senses. A combined (m) mind is created from Atman, which splits into the sky (space, or c^2) and Pran a (ε), which is surrounded by the mind. The reaction (impulse) creates knowledge (potential energy) that changes to mind (kinetic energy), which is a combination of thoughts. Mind (m) splits into Prana (ε) and space (c^2), and they recombine into mind (m). The creation can be represented as m = Prana/space = ε/c^2. This also means Pran strikes the space to create matter. Dualism is broken during this process. A triplet is formed from doublets. God is found everywhere in many multiplets. Atman is free from characteristics, and a singular entity can now be many and spread out everywhere. Some steps need to be followed. How can one become many by overcoming uncertainty (duality) and achieving three characteristics and many colors? Starting from the mind, a collection of thoughts can create a number from extremely high to low frequencies.

FIGURE 5.6
The dream of Vishnu gives an idea of a multiverse. Many cycles are created from one cycle. The god is dreaming about the creation. In his dream multiverse is created. Each universe is a bubble that contains the god himself accompanied by a god of creation emerging from the navel of the god. A goddess sits by the feat of the god. Inside the bubble, a few other bubbles can be seen. This creates a holographic world.

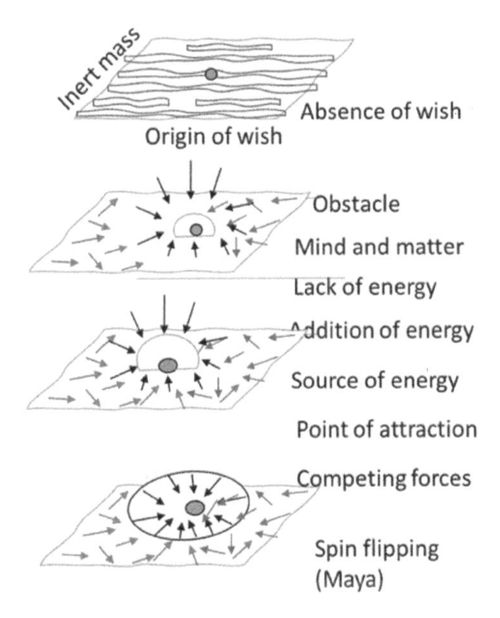

FIGURE 5.7
Creating curved space from a flat space in different stages. A point of strong attraction is created, which is countered by competing forces. An area is filled with wavy lines around a center. At some stage, two sets of spins in opposite directions try to compete for control of the center. Ultimately, the center is shielded by the spins.

Thought is converted into a condensed state, where very fine frequencies exist like small oscillations.

However, to reach this condensed state, emission and absorption of heat energy or heat loss have to be encountered. This heat exchange, e.g., emission, is manifested by the decrease of wave amplitudes as well as frequency. Low-frequency and low-amplitude waves exist in Atman as memory, which is omnipresent. Large waves excite our senses and creative impulses. While approaching our physical death, heat emission from the body creates condensation containing only the prime frequencies, which absorb all heat and can be released from the body. If heat remains in the body, then Atman cannot be created as a condensed state and stays in the body. We can compare this with water molecules that absorb heat and are vaporized. Atman comes back to Earth like raindrops. This can also be explained through the change of Atman! Initially, it is still and flat since space, time and space-time cause-and-effect (force) are not separated. However, small disturbances stop the formation of mass. Time leaves Atman in the form of energy (heat) or phase (magnetic field). Space in the form of a sphere is created as if Vishnu wakes up, which defeats all disturbances, and mass is created. However, again space becomes one static object like Brahma who represents Vishnu but a different entity of Atman (as the Sun is created from stardust). This is like shifting the center of the Universe to a local center (decentralized or from one Universe to another). Mass (or a large cluster of water molecules) can be moved from one place to another in space through a circular motion. In a vortex field, particles can move to a high altitude and also travel a large distance. In planetary motion, massive objects revolve around the Sun in an elliptical orbit, which is associated with angular momentum and the force of gravity.

Why is angular or rotational motion so important? It is the basis of our civilization. Wheels, pulleys, and gears give stability to a vehicle. It starts from a point and returns

FIGURE 5.8
One void can be filled with an infinite number of voids, creating a holographic world. A circle is packed with many small circles. Being inflated five petal-like structure is formed. Each petal contains three circles.

to the same point. Linear motion is not associated with an external force that is always present since the world is curved and no motion is free from force. The freefall of a particle from a height follows a straight line since gravitational force acts toward the center of the Earth. Any motion in a straight line not exactly normal to the surface of the Earth is curvilinear as a result of the applied force and the gravitational force. The criterion for the rotation of a particle is that the tangential velocity of the particle is always normal to the line joining the gap, i.e., the axis of rotation to the rotating particle. The rotation of an object is described by the centripetal and the centrifugal accelerations that act toward the center of rotation and outward, respectively. The criterion for the rotational motion is that the distance vector and the force are perpendicular to each other, forming a moment. The tendency for rotation is called the moment of a force that works in the perpendicular direction of the applied force. Rotation without translation can be achieved by the application of a pair of equal and opposite forces at two different points of an object. This combination of two moments around a fixed point is called a 'couple'. The operation of the wheel was invented based on the principle of rotational motion. The inertia (or mass) in linear motion can be replaced by the moment of inertia. The equivalent force in rotational motion is called 'torque', which is proportional to the centripetal acceleration pointing to the center of the motion and the moment of inertia. In the final part of this chapter, we unify the linear (along the axis) as well as the rotational motion in a vortex.

Vortex

In fluid dynamics, a vortex is a region in a fluid in which the flow revolves around an axis line, which may be straight or curved. Vortices form in stirred fluids and may be observed in smoke rings, whirlpools in the wake of a boat, and the winds surrounding a tropical cyclone, tornado, or dust devil. Vortices are a major component of a turbulent flow. The distribution of velocity, vorticity (the curl of flow velocity), as well as circulation is used to characterize vortices. In most vortices, the fluid flow velocity is greatest next to its axis and decreases in inverse proportion to the distance from the axis. In the absence of external forces, viscous friction within the fluid tends to organize the flow into a collection of irrotational vortices, possibly superimposed to larger-scale flows, including larger-scale vortices. Once formed, vortices can move, stretch, twist, and interact in complex ways. A moving vortex carries some angular and linear momentum, energy, and mass with it.

Vortex Geometry

In a stationary vortex, the typical streamline (a line that is everywhere tangent to the flow of the velocity vector) is a closed loop surrounding the axis and where each vortex line (a line that is everywhere tangent to the vorticity vector) is roughly parallel to the axis. A surface that is everywhere tangent to both flow velocity and vorticity is called a 'vortex tube'. Vortex tubes are nested around the axis of rotation. The axis itself is one of the vortex lines, a limiting case of a vortex tube with zero diameters. When vortices are made visible by smoke or ink trails, they may seem to have spiral pathlines or streamlines. However, this appearance is often an illusion, and the fluid particles are moving in closed paths. The spiral streaks that are taken to be streamlines are in fact clouds of the marker fluid that originally spanned several vortex tubes and were stretched into spiral shapes by nonuniform flow velocity distribution. Two or more vortices that are approximately parallel and circulating in the same direction will attract and eventually merge to form a single vortex whose circulation will equal the sum of the circulations of the constituent vortices. The motion described in some philosophical texts describing atoms can be found in a very similar description of a vortex field. We conclude this chapter with a description of a vortex that relates to the microscopic Universe.

Great Indian scholars described Brahmanda in two parts that resemble the atomic model. One is small like the nucleus of an atom (inner part). The other is the large (like short and long wavelengths) outer part that consists of five elements, all senses, and the sky (space). The inner part consists of the mind, heat, and knowledge. Brahmanda consists of Brahman (Shiva) and Sakti (Kali). Brahman is indivisible and a great zero (Mahasunya). It can split into many zeroes, i.e., Saktis like $0/n$ = zero, and are called Atman. They are many sparks that develop from one fire, like rays from the Sun, like stars from a big bang. Brahman is Purusha and can be divided into a large Brahmanda called 'Iswar' or 'Paramatma', or many small Brahmanda called 'Atman'. Three characteristics of Sakti are that Rajas gives repulsion, Tamas gives attraction, and Satwa creates equilibrium, which is the maximum energy. When the system is out of equilibrium, then the wave or radiation is emitted to compensate and bring back the system into equilibrium. Nonequilibrium starts with the formation of a dual state, which is converted into a triplet state. Their hypothesis can be explained in a system containing one atom (a zero) inside a resonator cavity (a large zero), forming a system of two concentric zeroes. Pran is like external energy that excites the atom. Atoms consist of a two-level system that has a ground and excited state, which is a signature of dualism or a doublet.

Now we would seek a three-body system that will also be treated as one, particularly at equilibrium. So we need two doublet states to achieve a triplet state. Unique Brahma is split as it wished (described as the Dream of Vishnu) into Shiva and Sakti, which is a transition of a single to a doublet and which, with the external energy, can be a three-state system. Duality is needed to create resonance. The resonance of the wave will be constant, at least for a long time against the nonresonant case where the energy will decay fast. Resonance is also a phase transition that has high intensity or the peak as if it is uplifting a body to the sky and does not return to Earth. Nonresonant is like recycling or rebirth, and the system must gain energy for resonance. Anyway, resonance needs two and only two objects: space and Prana created by dualism. It is like a bell that creates sound from the swing of the solid object in the bell, which is a resonator creating the space for the object to swing. This is also like a doublet. Now one needs to be many, at least two. So two bells, being coupled to each other, should ring and oscillate coherently, and that will create a compressed wave (bonding attraction), an expanded wave (repulsion), and a neutral wave (fundamental). This is a triplet state Tamas-Rajas-Satwa produced from two doublets. Similarly, an entangled system can be produced from many atoms, each having a doublet. God is everywhere in the same form. She lives actively in the heart of inactive Vishnu. She could be thought of as an external form in which fundamental frequencies arise. Sound is produced by her. She is Kali who created sound through her tongue and teeth in her mouth. The Mother Goddess is the very first vibration that produces light. Knowledge of fire produced in the brain (through mind–space interaction) comes out through her mouth like fire. She (the goddess) strikes Vishnu (sky/space) periodically to create a condensate. Mass is produced from condensation. This process looks like a rotational motion of some massive objects that leads to the formation of a vortex (field).

In Chapter 6, we shall explain the interaction of two vortices in the context of the quantum vortex. Arranging zeros or circular orbits to form a vortex structure is, I find, very similar to the model of Brahmanda.

6

Energy/Light and Space (Matter) Application in Condensed Matter Physics: Two Cycles

In *Vishnumaya* described in Chapter 3, we explained the process of how one can attain equilibrium through the interactions of a three-body system, e.g., Satwa, Rajas, and Tamas. They combine and form one unit. Here we also see three gods, Brahma, Vishnu, and Shiva (Kali), who combine their energy and form a very strong goddess, Goddess Durga, who can fight against an extremely evil force (Tamas). Such interaction is described in an ancient scripture (*Sri Sri Chandi*, described in Chapter 3) as follows.

Battle Between Mahishasura and Goddess Durga

Once upon a time, a fierce demon called the 'Mahishasura' was born from the unification of a buffalo (or Mahisha) and a demon (an asura) with superhuman powers. At that time, gods and demons were always fighting one another, and the gods usually won. Day and night Mahishasura was thinking about how to become more powerful than the gods and achieve *immortality*. He prayed to the god of creation, Brahma and received a boon that he could not be killed by gods or a man: *If he has to die, it should only be at the hands of a woman*. Being a gigantic demon, Mahishasura was confident that no woman would be strong enough to kill him. Mahishasura was invincible, and thus the inhabitants of the Earth were living in fear of the demons. After suppressing all mortals, Mahishasura decided to challenge the gods and finally attacked heaven, and the gods began to flee in despair from heaven.

The gods were in a pathetic state, wandering for years over fields and mountains. Tired of being in exile, they decided to consult the Trinity—Shiva, Vishnu, and Brahma—to come up with a way to destroy the demons so that they could return to heaven. *They decided to combine their powers to create one goddess*. The gods closed their eyes and began to concentrate all their thoughts on creating this invincible woman. Their divine powers and deep concentration worked, and soon a fiery pillar of light appeared in the sky. It was so bright that even the gods found it impossible to look at it. It was a mass of pure energy, produced from their combined power. From this energy, a goddess called Durga was created, who would be strong enough to vanquish Mahishasura. Shiva created Her face; Vishnu gave Her arms and Brahma provided Her with legs. All the gods presented Her with various weapons. In particular, Vishnu gave Her a special wheel-like device (Sudarshan Chakra), Shiva gave Her a trident, and the Sun god presented Her with His blinding rays.

DOI: 10.1201/9781003304814-7

FIGURE 6.1
A battle between the goddess Durga sitting on a lion and a buffalo-like demon and his associates. A ten handed goddess is riding a lion and fighting a half buffo like demon. She is carrying different weapons in her arms. A number of demons can be seen in the background.

Mahishasura attacked the goddess along with his other demon group and failed again to suppress the goddess. Mahishasura tried all the tricks he knew. He kept changing shape to confuse the goddess. From a man, he became a lion, then an elephant, but each time the goddess wounded him severely with Her weapons. The battle raged for nine days. Finally, the goddess killed the demon-king Mahishasura, who had taken the buffalo form again but was beheaded by the Goddess with the chakra that was given to Her by Vishnu. Thus Goddess Durga freed the world from the demon's tyranny. Indra and the other gods returned to heaven, and all were well and happy.

This story can be described as the interactions between particles. Mahishasura is a combination of a buffalo and a demon, hence representing duality (or a state of nonequilibrium) and disturbing goddess Durga. Durga, representing the state of equilibrium (or local point), destroys the duality. At this point, energy attains a maximum and destroys the noise produced in this process. It is like an extremely hot object dissolves all locally hot points. Or it can be a strong magnetic field that destroys the effects of other small magnets. The nature of nonequilibrium is always changing with time. Due to instability, the phase of the nonequilibrium point changes as seen in the changeable form of Mahishasura, who is destroyed by Goddess Durga. Finally, Durga (Satwa), Her carrier the Lion (Rajas), and the defeated Mahishasura (Tamas) form equilibrium. Goddess Durga, having ten hands, represents the Sun, which spreads out light in all (ten) directions. Mahishasura may represent clouds in the sky having changeable shapes and sizes (and colors), which prevents the Sun from sending light to the Earth to produce life (the lion). Mahishasura acts like a magnet that has two poles or a dual character that can work against a nonpolar system like a Cooper pair in a superconductor with a spin-zero state (will be discussed in Chapters 7 and 8). It forms a degenerate state that stops the tunneling of Cooper pairs. A superconductor rejects the incoming magnetic field since it produces its own magnetic field (Tamas). However, by increasing the external magnetic field (or an electric field) and temperature the superconducting state can be destroyed. The superconductor becomes a normal conductor.

We have gone through the era of Newton and Leibnitz in classical mechanics, and now we extend our discussion related to the modern-day space—time (and gravitation) concept (as described by Einstein in curved space). The equilibrium model can be applied to the magnetic fields associated with rotation and angular momentum. A generalized model of rotational motion can be applied to vortex structures. Curvilinear motion can be applied to the theory of electromagnetic interactions. The magnetic field lines are curvilinear, which emit from one pole of the magnet and merge to the opposite pole, hence forming a closed loop. Two magnets can be coupled through the magnetic field lines, which also form a closed loop. This model can be generalized by a vortex, which can also be found in magnetic and superconducting materials. The interaction of charged particles (electric fields) and spin ensembles (magnetic fields) can be generally understood in a vortex structure. This concept will be compared to the philosophical ideas of a vortex or holographic Universe, which is addressed in the evolution theory of Sankhya. In electromagnetic theory, electric and magnetic fields are mutually orthogonal fields that can be represented by a dualistic view as the Rajas (a repulsive force) and the Tamas (an attractive force). By combing these two fields, light (a neutral field) can be produced as the Satwa characteristics. To describe the vortex model, we should understand the electromagnetic theory. In Chapter 5, we described the mass or inertia and the attractive force, i.e., gravitation. Here we deal with the massless particles (i.e., light quanta or photon) or spins (the phase). We shall explain Faraday–Maxwell's laws of electromagnetism and the basic properties of a magnet. One of the most interesting parts of the Sankhya philosophy is the collection of mind or intelligence, which is the first manifestation of Prakriti. This is transformed into the 'ego' (or soul), which is dissociated into five work-related senses and five knowledge-related senses and one mind. This description can be applied to space–time distortion (Chapter 9) ideally to the vortex and the holographic Universe. To understand this model, we should review the basic concepts of electromagnetism.

Electromagnetism: Faraday–Maxwell

Faraday's law of electromagnetic induction is a basic law of electromagnetism predicting how a magnetic field will interact with an electric circuit to produce an electromotive force—a phenomenon known as electromagnetic induction. Faraday's law of electromagnetic induction says that, when a magnetic field changes, it causes a voltage, a difference in the electric potential that can make electric currents flow.

According to Faraday's first law, any changes in the magnetic field of a coil of wire will cause an emf to be induced in the coil. If the conductor circuit is closed, the current will also circulate through the circuit, and this current is called induced current. There are methods to change the magnetic field: (1) by moving a magnet toward or away from the coil, (2) by moving the coil into or out of the magnetic field, (3) by changing the area of a coil placed in the magnetic field, or (4) by rotating the coil relative to the magnet.

According to Faraday's second law, the magnitude of emf induced in the coil is equal to the rate of change of flux that links with the coil. The flux linkage of the coil is the product of the number of turns in the coil and the flux associated with the coil. From the bases of electromagnetic theory, Faraday's idea of lines of force is used in well-known Maxwell's equations; e.g., Faraday's law change in magnetic field gives rise to a change in electric field, and the converse of this is used in Maxwell's equations. We explain these features based on the interference pattern arising from two competing objects.

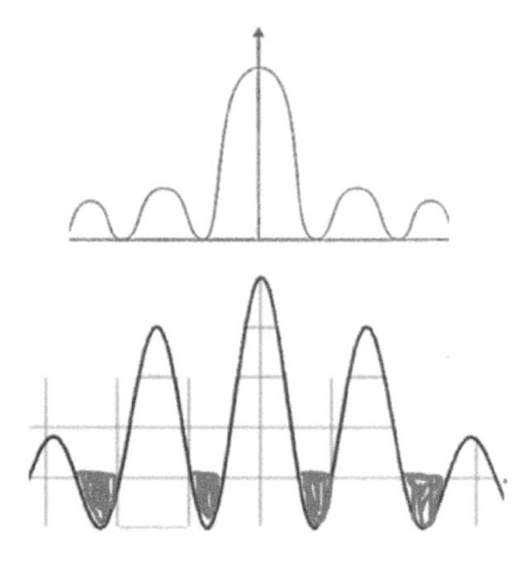

FIGURE 6.2
A wave consisting of a central maximum and several minima. An interference pattern is formed from the competition between the positive and negative parts of the space. This is an example of duality that also shows a central bright peak.

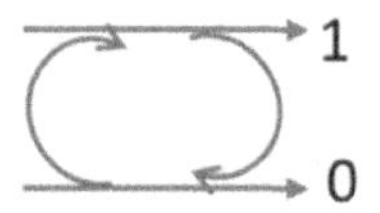

FIGURE 6.3
A two-level system showing two levels representing the ground and an excited state which are connected by two oppositely rotating curved arrows. The interference pattern can be compared to the battle of demons with the god(s)—a dualism to nondualism transition.

In the previous chapter, we described static equilibrium and the Sankhya philosophy. Here we describe the motion of the dynamics of a particle which can be treated by circular motion. The curvilinear motion can be described by Newtonian mechanics, which explains planetary motion. We describe the dynamic equilibrium based on the Sankhya philosophy, which also gives a general understanding of Newtonian mechanics and three-body interaction. Sankhya's philosophy describes creation as a combination of three different characteristics that are also attributed to three forms of deities (god or goddess). This is the foundation of classical mechanics where equilibrium is attained: three in one—One egg and two birds or one bird and two eggs. One Purusha and one Prakriti in combination form a circular orbit system. Sunrises and sunsets suggest the rotation of the Earth. The change of phase of the Moon also shows daily rotation around a center. But the revolution part is missing, which adds another center, and the orbit becomes elliptical. One elliptical orbit holds both the rotation and the revolution of celestial bodies. Two different rotations can be described as one being static and the other being relatively dynamic. All

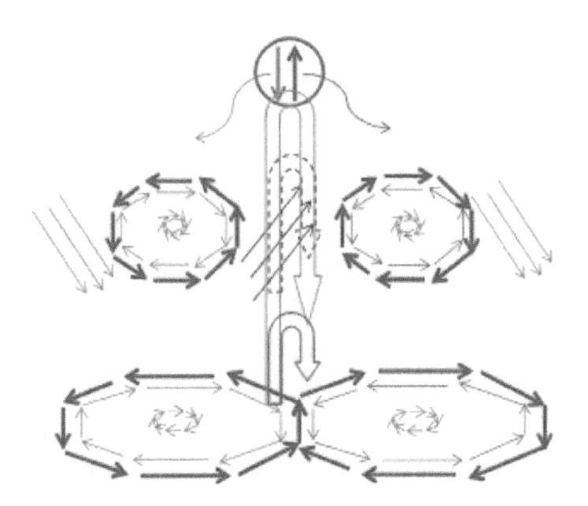

FIGURE 6.4
Two-dimensional projections of two interacting vortices. Each vortices have two circulating currents in opposite directions. A set of two loops rotating in opposite directions is shown by connecting spins. Each loop is associated with an inner circle which rotates in opposite direction of the outer circle. These loops are connected by a feedback loop which grows with the size of the loops.

motion is relative. The Sun as a static object can see the Earth and Moon; one is static and the other dynamic. So we need two poles, one static and one dynamic. A supreme god can hold both static and dynamic phases. Diagrammatically, He lives at the apex of a triangle that controls the other points of the triangle. Relative changes of these two points are like the foci of an ellipse and the position of the Sun, and planets revolve around these two centers in an elliptical path, which was experimentally observed much later than the philosophical model. For example, Mahavishnu (the great Vishnu) holds both Vishnu and Laxmi. Similarly, we see gods described as Sadashiva (the highest manifestation of Shiva), who holds both Shiva and Shakti. They describe three planes of Atman; one is constantly changing or transforming, one is not changing (static), and the third is the superposition of both changing and static phases. He is playing with living and death.

We have gone through the era of Newton and Leibnitz in classical mechanics, and now we extend our discussion related to the modern-day space–time (and gravitation) concept (as described by Einstein in curved space). The equilibrium model can be applied to the magnetic fields associated with rotation and angular momentum. A generalized model of rotational motion can be applied to vortex structures. Curvilinear motion can be applied to the theory of electromagnetic interactions. Magnetic field lines are curvilinear, which emit from one pole of the magnet and merge to the opposite pole, forming a closed loop. Two magnets can be coupled through the magnetic field lines, which also form a closed loop.

Magnetism

For a long time, some of the basic properties of magnets found in nature were known. For example, when two magnets or magnetic objects are close to each other, a force attracts the poles together. Magnets strongly attract ferromagnetic materials such as iron, nickel, and cobalt. On the other hand, when two magnetic objects have like poles, called the 'north' (N) and 'south' (S) poles, facing each other, the magnetic force pushes them apart. Modern science describes magnetism caused by the motion of electric charges. Every substance is made up of tiny units called 'atoms'. Each atom has electrons, particles that carry electric charges. A magnetic field consists of imaginary lines of flux coming from moving or spinning electrically charged particles; examples include the spin of a proton and the motion of electrons through a wire in an electric circuit. What exactly a magnetic field consists of is somewhat of a mystery, but we do know that it is a special property of space. The lines of magnetic flux flow from one end of the object to the other. By convention, one end of the magnetic object is called the 'N' or north-seeking pole and the other the 'S' or south-seeking pole, as related to Earth's north and south magnetic poles. Magnetic flux is determined by this movement from N to S. The Earth does not follow the magnetic configuration in the illustration; instead, the lines of flux are opposite those a moving charged particle. Although individual particles such as electrons have magnetic fields, larger objects such as a piece of iron can also have a magnetic field, as a sum of the fields of its particles. If a larger object exhibits a sufficiently great magnetic field, it is called a 'magnet'. The magnetic field of an object can create a magnetic force on other objects with magnetic fields—we call that force 'magnetism'. When a magnetic field is applied to a moving electric charge, such as a moving proton or the electric current in a wire, the force on the charge is called the 'Lorentz force'. If the total number of lines of force comprising the magnetic field is called the 'magnetic flux', the magnetic flux density is defined as the flux passing per unit area within a material through a plane at right angles to the flux. The intensity or strength of magnetic field at any point within a field is measured by the magnitude of a force experienced by the unit N pole placed at that point. This intensity of the magnetic field has a definite direction at every point acting along the lines, and these points are called the 'magnetic lines of force'.

This model can be generalized by a vortex, which can also be found in magnetic and superconducting materials. The interaction of charged particles (electric fields) and spin ensembles (magnetic fields) can be generally understood in a vortex structure, which is

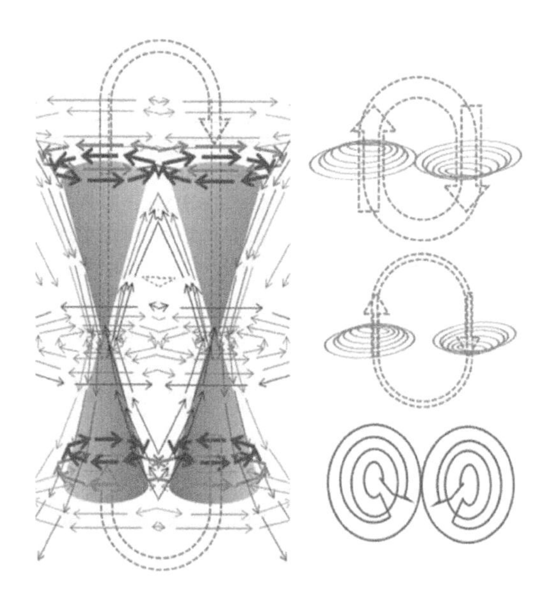

FIGURE 6.5
A set of two cones are connected at their tapered ends. They are surrounded by circulating spin currents around the two junctions. These cones are connected by two curves arrows which represent a feedback loop. They represent two interacting vortices to form a feedback loop. They can represent a vortex anti-vortex transition. A spin-triplet system is formed.

described in this chapter. This concept will be compared to the philosophical ideas of a vortex or holographic Universe, which is addressed in the evolution theory of Sankhya philosophy.

The Soul or Atman moves like magnetic field lines in the body like elementary particles, electrons, protons, and neutrons, and our mind is divided into thought, intelligence, and consciousness. Our mind behaves like the revolution of the Moon around the Earth in different phases. Ultimately, the rotation and revolution of a planet are the basis of all meditation. The Sun, as the ultimate constant, is not associated with a decay of energy or darkness. The Sun emits different kinds of radiation; however, a single model is the magnetic pole structure of the Universe that emits field lines in all directions. By that, I mean the four directions like the four heads of Brahma. In a two-dimensional plane, the lines look aligned in left and right directions with a line in a straight upward or downward direction. This is the model describing the Holy Trinity that can explain the creation process of matter. Moreover, two coupled magnetic centers will make the picture of the vortex-like Universe (related to space–time distortion, described in Chapter 9) more complete. The field lines emitted from one magnet enter the other magnet. Hence, a closed world is made by coupling a large number of magnets. Ultimately, the Sun produces magnetic fields, which is *Maya* different from that of electric fields or light or heat. The concept of a magnetic field is widely accepted in deities describing the supreme god- or man-like Purusha accompanied by two wives (or two assistants). The central part of the deities that shows the union of Purusha-Prakriti (like Shiva-Parvati or Krishna-Radha) can be compared to the north and south poles of a magnet. The same model can be applied to the electrical fields. This structure of a vortex or vortices can explain most of the exotic transport features in

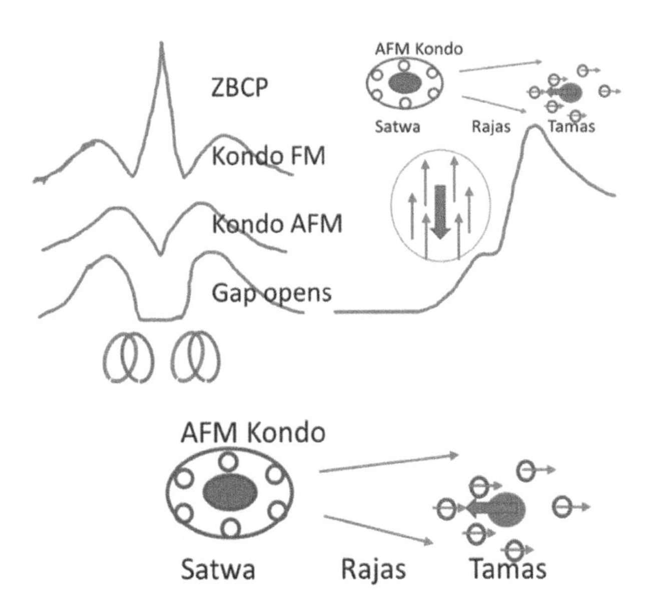

FIGURE 6.6
Unconventional superconductivity is produced from the spin-flip, which exhibits a resistance peak before reaching zero and the finite density of states in the superconducting gap. The conductance spectrum has a central peak and two smaller peaks on each side. Under some conditions the central peak becomes a minimum and then a gap is created. This transition corresponds to a screened charge where a filled circle is surrounded by a few unfilled circles which interact at higher temperatures. The temperature dependent conductance curve shows a hump at a particular temperature.

condensed matter physics. A model of the holographic Universe can be described by the vortex fields (which also described the magnetic fields in the human body). An application of Sankhya philosophy can be found in Goddess Saptasati or *Sri Sri Chandi* (as a part of Atharva Veda). This book talks about overcoming bondage or knots of life or the complexity of space. Let me repeat what is being said. Goddess Kali (introduced in previous Chapters says, "I am your mother (Shakti) who initiates creation". If She provides a strong force, then light can be captured, and due to disorder or multidimensional space, it can reflect in different paths and cannot escape from a space like a black hole singularity.

A similar effect can be seen associated with dualism. A light particle can be deflected either to the positive or negative side in a space (holographic) of dual nature created by positive (+) or negative (–) charge or N or S type of magnetic poles. Brahma can do so by overcoming all four directions through the neutral point. But the light in a funnel-like space moves in a spiral and cannot be liberated in the dualistic space. So noise will be created, but She will cut all knots for you to be liberated. She also has four hands holding or performing four different (opposing) actions. The first section, overcoming the knot related to Brahma, shows Goddess Kali as a very strong introvert force that produces Earth from cosmic dust or plasma-like water. The main obstacle was Madhu-Kaitava (described in Chapter 3) who stopped the energy to fall onto Earth from the Sun. The second part is Vishnu, where the goddess is created from the Sun by concentrating his power to a point to create the Moon. The Moon was not stable, the phase was changing, and it was described by the demon Mahishasura, who changed from one animal to another. The third was associated with Rudra for the creation of the Sun itself. I am trying to explain only the first part of the book since the remaining part of the book remains largely unclear. Starting from the first knot, we say that She breaks 0s or all small cycles and small oscillations of individual atoms and finally creates a collective state of motion. It is like two wheels that produce a gear system. The big one rotates, and the small one assists the motion or feedback. The big wheel is like a large oscillator, and the small wheel is a small oscillator. They produce big or classical waves and small waves or quantum fluctuations, respectively. The small wheel does not want to move due to inertia, i.e., it moves back and forth around its axis. Their action is described by uncertainty, which arises from duality. The big wheel wants to break this duality, so it pulls the connection or belt. The attraction becomes strong in the same way as She creates very strong attractive forces to bring everything together toward her. The small circles can shrink if a field is applied. It becomes smaller and smaller so that it cannot apply any resistance practically to the big wheel. The small circle can form like a vortex by applying a field from the big circle like a magnetic field. This is like forming raindrops from a large reservoir of water, i.e., a cloud. It can ultimately overcome duality. Or She can add more mass to the cloud with more vapor so that it cannot hold itself and breaks down into small fragments. The cloud is opposing the water or a big cycle. The cloud is a small cycle that should be destroyed since it produces duality. By applying a negative charge to it, the cloud can be pulled down to Earth and produce rain. It can be broken by a field, e.g., a strong electrostatic field such as lightning. We must break the duality within a very, very short time; otherwise the mass will relax and produce duality. Lightning is a process that works in no time and creates polarity. Time is the best weapon, which is represented as a short time pulse (Δt) called 'Kaal'. It breaks up everything from velocity to phase. It is like knowledge or mind that works in no time. Kaal is the smallest period you can imagine: $F = mdv/dt$ where dt makes up the transformation, which changes the inertia. The change of time is $dt = m\, dv/F$ and is given as the ratio of change of momentum and the applied force. From $\nabla \times E = -dB/dt$, we get $dt = -dB/(\nabla^* E)$ which is the ratio of change of magnetic field (B) and curl of electric field (E). Kaal involves all mind, life, and knowledge. dt determines how this transformation happens. 'dt' is the present time that cannot be stored. It always progresses: (1) dv/dt breaks inertia d/dt (change of velocity); (2) $d\varphi/dt$ (Faraday) > breaking phase (φ); (3) breaking all satisfaction or differentiation on Maya, which creates a state of no attraction or repulsion and attains a constant velocity. Let me compare these concepts with the Faraday–Maxwell laws of electromagnetism.

Maxwell's laws are based on a set of coupled partial differential equations that, together with the Lorentz force law, form the foundation of classical electromagnetism, classical optics, and electric circuits. They describe how electric and magnetic fields are generated by charges, currents, and changes in the fields. Maxwell's equations can also be formulated on space–time like that of Minkowski space where space and time are treated on an equal footing. The direct space–time formulations make manifest that Maxwell's equations are relativistically invariant. Because of this symmetry, the electric and magnetic fields are treated on an equal footing and are recognized as components of the Faraday tensor.

Maxwell's first equation signifies that the total electric displacement (D) through the surface enclosing a volume is equal to the total change (ρ) within the volume: $\nabla D = \rho$.

Maxwell's second equation signifies that the total outward flux of magnetic induction (B) through any closed surface is equal to zero: $\nabla B = 0$.

Maxwell's third equation signifies that the electromotive force (E) around a closed path is equal to the negative rate (t) of change of magnetic flux linked with the associated path: $\nabla \times E = -dB/dt$.

Maxwell's fourth equation signifies that the magnetomotive force (H) around a closed path is equal to the conduction current (J) plus displacement current through any surface bounded by the path: $\nabla \times H = dD/dt + J$.

An important consequence of Maxwell's equations is that they demonstrate how fluctuating electric and magnetic fields propagate at a constant speed (c) in a vacuum. The microscopic equations have universal applicability but are unwieldy for common calculations. They relate the electric and magnetic fields to total charge and total current, including the complicated charges and currents in materials at the atomic scale. The macroscopic nature defines two new auxiliary fields that describe the large-scale behavior of matter without having to consider atomic-scale charges and quantum phenomena like spins.

Maxwell's equations in curved space–time, commonly used in high-energy and gravitational physics, are compatible with general relativity. Albert Einstein developed special and general relativity to accommodate the invariant speed of light, a consequence of Maxwell's equations, with the principle that only relative movement has physical consequences. The publication of the equations marked the unification of a theory for previously separated described phenomena: magnetism, electricity, light, and associated radiation.

We can assign Satwa-Rajas-Tamas to different physical quantities. Here $B\uparrow$ and $B\downarrow$ represent the Gods and the demons, respectively. Gods are forming large cycles, and demons are forming small cycles: small B vs. large B. The demon Mahishasur is a Rajas or Tamas working against She, which is life and also a Rajas. Small cycles work in opposite directions to big cycles. Inertia arises from the small rotation that works against the big cycle. The big wheel is the creator (Purusha) or the Goddess (Shakti). Small wheels are demons or reactions. Both Vishnu and demons produced from him are inactive or Tamas, i.e., inertia which is broken by mass.

The Kaal (dt or Δt) or the duration of the pulse splits everything. In Chapter 1, we said the Universe is Brahmanda (which is Brahma Anda) and that 'Brahma' means spread out. 'Bishtar' in Sanskrit means that which progresses with time. An example is Agni, which always progresses. 'Anda' means egg, a seed of creation, an oval-shaped body that contains mass and energy together, like a natural egg containing a red core surrounded by

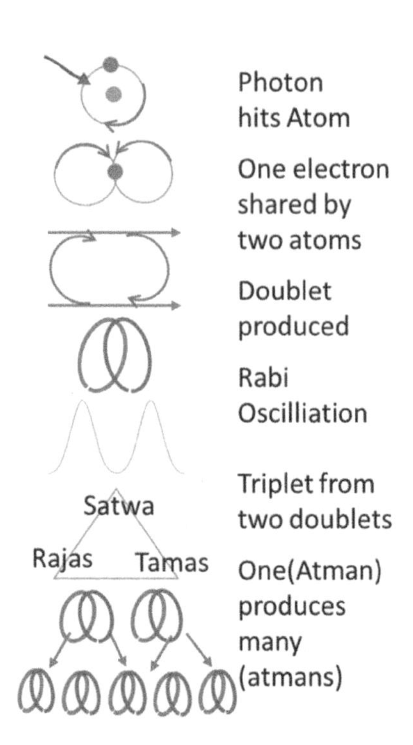

FIGURE 6.7
A singlet state is split and forms a doublet state. Two doublet states recombine to form a spin-triplet state. This process finds similarity with the trinity described in Eastern philosophy. Two sets of oppositely rotating circles are linked with an arc shaped arrow representing a feedback process. (bottom) An incident photon strikes a circle like atom and splits it into two opposite rotating circles. I two level system is formed. This is transformed into a three-state system. That is represented by a triangle having three apexes corresponding to three different characteristics. Finally, a multilevel system is created.

water (the white part) and covered by a hard shell, the skin of the egg. So the Universe is expanding in the form of an egg. An egg has a generic shape like our head, like the womb of a woman, and like other reproductive organs even for men, and, most importantly, like our eyes including eyeballs. It is also associated with other sense organs: ear, nose, tongue, and skin, which have pores through them so that exchange of air or information takes place. Therefore, an exchange interaction is a cause for having the holes, and two holes (or more) are always necessary for the exchange that follows dualism. One can treat them as input and output, which result from attractive and repulsive forces. Regarding image, the creation dualism model plays a crucial role, particularly in self-representation. We are attracted to our image through love, but we cannot merge into our image initially. So we follow our image hoping that we merge. The acts of following my image and my image following me create a process called 'moment', which is associated only with angular motion, i.e., with velocity in only a curved space. Catching our image without merging produces a shape that is either a circle or a sphere in three dimensions. This is Anda, or 'seed' or egg, which grows up with the addition of energy or shifting away from equilibrium. In modern (Newtonian) science, this circular motion continues as long as an equilibrium of force is maintained. The forces are centripetal (toward the center) and centrifugal (off centered). These may be called attractive and repulsive forces, which work together against each other and are defined as Tamas and Rajas. The center, called

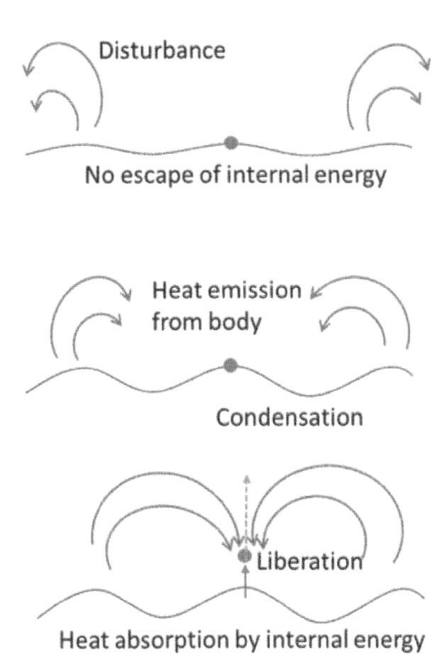

FIGURE 6.8
The superconducting gap is compared with the ground state or a state of deep sleep. However, disturbances from the states outside the gap can excite the gap state, and a finite density of state can be created in the zero-energy region, which exhibits the zero-bias conductance peak. Two oppositely oriented curved arrows originate around a central point. (middle) At some stage they try to control the center. (bottom) These arrows try to merge at the center which can release energy.

'Satwa', remains static, which keeps the balance between Rajas and Tamas. The whole process creates a zero or Maya. Many zeroes forming a 'garland-like' shape is Mahamaya or a great illusion. She can be defined by many like Shakti, Vishnumaya, intellect, sleep, night/darkness, brilliance, etc. One of them is 'Shadow', her own shadow, and this gives a self–spin, a perpetual motion. She wishes to split herself into her own shadow, which is like a conformal shape. In the ancient text, Sri Sri Chandi goddess is defined as 'Ardha Matra', or half of three elements. Three elements form a full entity (body) and half create a duality. It is like a half flux where the first and second half are in antiphase. So, from the full circle, two semicircles are produced, which form even and odd functions like women- or man-like features. Then the exchange interaction can take place. Mother Durga, who is white, created the black goddess Kali as the representation of darkness, or Tamas. She came out from the body of Vishnu or Shiva or even Durga as Shakti. Ultimately, a state that is a combination of Shiva and Shakti, a superposition of two states, is created. They play with each other and form a sphere called a 'Bloch sphere' in quantum science (discussed in Chapters 9 and 10). This sphere, or 'Anda', is the most fundamental unit of the Universe. Even the flat space can be considered a part of the sphere. In practice during the creation of the sphere, other small circles can be formed due to the imperfection or defects, which should be destroyed. It is the world of the goddess who wishes to have a world of harmony (one main orbit and the other smaller orbits around her). She kills all other dualities that are represented by demons in pairs like Madhu-Kaitava (attractive and repulsive forces), Mahishasur (changing phase of different animals), Chanda-Munda (spheres made of wish and mind that are unstable), and Shumbha-Nishumbha (holes made of good and bad things comes from Kumbha, which means vortex or antivortex). She does not want them to form a complete cycle or sustainable dual characters and therefore breaks the cycle halfway. A full cycle will not carry any mass, and a half cycle will. So masses are produced based on the attractive part of the force. As deadheads, these demons are associated with the goddess Kali as a garland where the heads can be considered as a planetary system. One main orbit of the string is decorated with small orbits or many garlands. This is the only model of the Universe we follow.

In the ancient text, goddess Kali says:

> I am the one the only circle or great zero. From me, other zeroes are produced like Vishnu, Shiva [or Kali], and, finally, Brahma from the interaction of initially inactive Vishnu [Tamas] activated [by Shakti/Kali]. I split the force into attractive [Kali] and repulsive [Vishnu] or vice versa and produced the equilibrium Brahma. I produced mass from other attractive and repulsive forces, Madhu Kaitava through one of my children Vishnu, so another triangle [three elements] around Vishnu with Madhu Kaitava is formed. A chain reaction begins. Similarly, my offspring Durga and Kali have created from the combination of different [three] elements. Each time a radiant sphere [a Brahmanda] is formed, that also poses a dual character and then gives three elements. The holy triangle is split into two halves. In this way, I create thousands of zeroes or circles. So I have thousands of heads/eyes/hands and legs. I am like a lotus since I can create a gap between the petals by using my repulsive force, and at the same time, I put them together using my attractive force. Opposite forces like electric and magnetic forces create light. I create the space. I have both inertia [Tamas], change of velocity, and knowledge, i.e., Satwa. I have knowledge of inertia and space, which interact with each other. I am the one who controls the tunneling process. I am Brahmanda, which is a sphere and cannot pass through a small hole/space. I can be blocked by an obstacle that is inactive [like Vishnu]. So I split myself into concentric spheres. I create space between these spheres. I create many zeroes from one. [This is like splitting a fermi sphere (full of charge) or a cloud ball containing charge into different spheres by applying a magnetic field.] So I look like a magnetic field that creates a rotation of Shiva [in Daksha Jagga described at the beginning of Chapter 5] and also that splits my body using his [own] magnetic field. Anyway, I maintain a dual character, i.e., two concentric circles move in opposite directions. They are neither attracting nor repelling each other. They continue this motion or desire-less interaction throughout time. This is called Karma [action], which does not deserve any profit or energy release. [This works like a Cooperon or a binary star, where two particles (or even massive objects) of opposite characters circle each other.] (See Figure 6.8.)

However, if any imperfection or deflection arises, we have to change this motion to counteract the local disturbances. So the circles change to destroy the local circles based on the third element, which is a combination of attractive–repulsive forces and inertia (or knowledge). At the bottom of the tunnel or funnel or vortex, we see *Karma* or the duality but outside the singularity (which is also dual newly) the forces arise, and three elements are formed. In the reverse order of three elements, singularity and duality are formed. All forces and energies focus on a single point, the dualism is destroyed, and heat or light is produced. An ultimate zero is formed from a big zero or many zeros. After reaching the singularity, an explosion takes place with sound and light or forces diverging. The 'bang' represents the change of properties of particles in terms of charge, spin, phase, etc. so that an opposite trend starts (a flip). This will be elaborated on in the context of condensed matter in the next chapter.

From the ancient text, we repeatedly see a flip of spins or rotations as the trick used to kill demons. This is a kind of the greatest property of Vishnu as described in Chapter 3. This flip is Mahamaya, a great illusion that is described in detail in this chapter focused on electromagnetism. Sometimes they are locked alone and do not flip, and they spin in an upward or downward direction and oppose each other. This situation (Kondo) can be overcome by spin flip which goes to a superconducting state (see Chapter 8). Nuclear reactions create a gap (space). Between electrons and protons, a neutron comes. It forms an $e^-$and p^+ from $n \rightarrow e + p$. Brahma cannot cross the energy gap until sound (the goddess Saraswati) adds energy to an electron. This consort of Brahma and goddess Saraswati release an electron across the gap. This is called 'liberation'. The filled state wants to unite with the unfilled state, which gives liberation. Goddess Saraswati is described as the union of three elements that she holds in her hand, i.e., a musical instrument, a red object, and a book that signifies work (Rajas) and knowledge (Satwa). She connects filled and unfilled states (duality) and wears a garland made of white stones. This is nothing but a model of the Universe having many planets and multiresonance. Multiple transitions between filled and unfilled states take place, which creates multiresonance or deep sound. Between the insulator and conductor, the gap is filled by a weak or strong coupling liquid or field. This liquid is like nectar, which can be described as a disordered solid. Saraswati is the goddess of this high-quality liquid. It can be ejected from the space between the filled to unfilled state once or twice. The quality of the fluid is the creator (Brahma) can make a reentrance called his 'rebirth'. For a rebirth, we need two vortices. One object's field is received by another. The liquid (or field) comes back to the second from the first one, completing the cycle. It is like two magnets lying in parallel with opposite polarity so that the flux lines of one magnet enter the other end and vice versa. This is like the Earth and Moon, which are considered magnets that are tied with their flux lines. They love each other by sharing the flux. This is an exchange interaction. One circle becomes two and many by sharing a common object. The Sun is tied to the other planets by sharing gravitational fields (or magnetic fields). This basically unites two magnets, i.e., four faces as Brahma has. The knowledge is divided into four poles N, S, S, and N. The connecting point where two circles of V/AV meet gives the quantum critical point. It is a point without dimension (such as the Dirac point). Vishnu kills Madhu-Kaitava at this point, which is not filled or unfilled, i.e., the point does not belong to the sea or sky. All noise merges at this critical point and is converted into mass when they move to either side or band. Vishnu divides massless particles into e^- or h^+ at this point and adds masses to them. Madhu-Kaitava is a cluster that forms a strong collective motion or boson (Vishnu is also a boson). The rise of Vishnu is the creation of the Fermi level based on a repulsive force of the goddess (Maya or Shakti) of heat. Brahma can be described as 2 photons that emit from the Fermi Sea of Vishnu, which can be the emitter with two polarities (four heads in total) if the gap is created by the strong repulsive force of the goddess (Maya-Shakti) by heat or a magnetic field. Madhu-Kaitava otherwise forms a condensed state that stops the formation of an energy gap or a massive particle. Now we shall compare the concept of the Sankhya philosophy with the model of the following vortex.

Quantum Vortex

A vortex is simply a decoupling of a plane from space into two or more planes through the creation of imaginary space in between. Vortices are created from multiple reflections or retroreflection, allowing for time-reversal symmetry breaking. Vortices cause spin-charge separation by creating space at their core. Initially, before the formation of the vortex, the space is isotropic and spherical, but then the symmetry is broken two times: first along with the core and, second, normal to the initial breaking. This is ultimately like a tube being split into two halves, and then each half is further divided into four halves. Each half maintains a $1/2\pi$ phase as it wants to catch its neighboring halves. Some eminent theoreticians proposed a suitable model to explain how two vortices interact with each other in a medium such as a spin liquid. This model shows that a bosonic liquid placed in a potential initially maintains time-reversed symmetry. Due to this nature, the system becomes 'locked' like an insulating state. To unlock this system, one has to apply a π-pulse to one of the components to cause the time-reversal symmetry to break. This is implemented by placing a barrier or a hole-like state. This is equivalent to an AB ring (or cylinder) where spin and charge are confined and a spinful excitation orbits an $hc/2e$ vortex. The Chern–Simons gauge field effectively attaches flux tubes of half strength to each spinon and vortex. This can be formally obtained by introducing branch cuts emanating from every vortex, across which fermionic wavefunctions must change signs. Each of the hc/e vortexes is associated with one of the spin currents. We expect such a double vortex condensate to appear in the dual description as a condensate with a doubled dual charge and have a half dual flux quantum. This looks like two magnets or two dipolar molecules in opposite orientations but connected with flux lines. Two hc/e vortices aligned on the x-axis and two other hc/e vortices aligned on the y-axis form a quadrupole. This would look like a quadrupolar structure of a d- wave or order parameter. To construct a vortex, we can combine two AB rings (with two triangles and tunnel junctions). In practice, the AB rings were constructed from multiple tunnel junctions as shown in our previous report. Vortices are more universal and can work like magnetic cores within the superfluid; however, they can drift, split (bifurcate), condense, be braided together, and annihilate. The spin-charge separation model or the holon–spinon model could not explain triplet transmission in the superconducting spin valves (at least this field theory model was not used, and it did not connect any other known interactions such as Rashba spin-orbit coupling or the Kondo resonance).

Now vortices have the potential to describe certain interesting phenomena such as weak localization (WL), weak antilocalization (WAL), Aharonov–Bohm (AB) oscillations, spin-orbit coupling (SOC), and the Kondo effect. Weak localization is a phenomenon found when an electron travels through a weakly disordered medium. This results in a backscattering of the electron as well as the forward scattering. Both paths can interfere constructively or destructively. The Kondo effect is characterized by the gradual flipping of a spin in a spin-triplet and happens because of the Kosterlitz–Thoules (KT) transformation. Rashba spin-orbit coupling occurs when the spins are initially confined in the vortices, coupled with different levels such as S or F sublevels, and this allows the transfer of energy. Based on the experimental observation of the reentrant superconductivity magnetoresistance (MR) peak, the zero-bias conductance peak, and the spin-valve effect in a disordered superconductor or a spin-valve structure we suggest a phenomenological model for a general understanding. Features of recent spin-triplet structures can be explained by a generalized dual vortex theory without the effect of a well-defined magnetic layer. This general approach can explain the formation of triplet superconductivity from the superposition of dual vortex fields and two spin currents (Figure 6.6).

Among all forms of dualism, concepts used in physics electric–magnetic duality is the most famous one where electric (B) and magnetic fields are related to the symmetry through the Faraday–Maxwell equations $\nabla * B = \frac{dE}{dt}, \nabla \times E = -\frac{dB}{dt}$. The symmetry

$E \rightarrow B$ *ad* $B \rightarrow -E$ is known as duality, which, although it still holds in the presence of charge and currents, finding such symmetry between electric and magnetic charges in quantum field theory was difficult. Later a symmetry was suggested that would exchange electric and magnetic charges but that must exchange the quantum of electric charge with a multiple of the quantum of magnetic charge. This symmetry has appeared like Faraday–Maxwell's equations, where the symmetry must exchange elementary quanta with collective excitations since, for weak coupling, electric charges arise as elementary quanta and magnetic charges arise as collective excitations.

This model was applied to the condensed matter as vortex–boson duality. However, a precise formulation for this duality is lacking. Vortex–antivortex interactions were shown through the Kosterlitz–Thoules transition in 1+1 dimensions; however, this is not firmly established in 3+1 or higher dimensions (2+1*d* is still being developed). It is generally suggested that a vortex interacts with an antivortex via a spin-wave fluctuation in a superfluid. The vortices act as sources and sinks of supercurrents, and therefore supercurrent is no longer in the vortex condensate. Models claiming charge–vortex duality based on a statistical transmutation of electrons involving the binding of electrons to magnetic fluxes or vortices was suggested in the frame of *anyon* superconductivity. A formal duality transformation between particles and vortices forms a two-dimensional Bose system that was believed to explain the origin of high-temperature superconductivity.

Summary

In this chapter we explained one of the most important similarities between the theory of electromagnetism (electric and magnetic fields) and the dualism in philosophy. We begin with a story describing the interactions between the good and the evil forces given in an ancient Hindu scripture. One goddess is fighting two demons representing duality. We shall extend this story in the next chapter again. We have presented a quantum vortex model as an extension of the classical model which will also be extended in Chapter 7.

7

Condensed Matter Physics—Creation of Particles—Bosonic vs. Fermionic: One and a Half Cycle

- **Battle of the particles:** Boson vs. fermions and Pauli's exclusion principle
- **Steady-state of a quantum fluid:** Bose–Einstein condensation
- **Uncertainty principle**
- **Quantum vortex**
- **Spin-orbit coupling**

Story of the Battle between Goddess Kali and Two Demons; Chanda-Munda

In this story, we find one mighty goddess (Durga) transformed into another very forceful goddess (Kali) in order to fight two pairs of demons. The two demons, Chanda and Munda, were the assistants of the demon-kings Sumbha and Nisumbha. The propitiated Brahma, through intense austerities, gained the boon to the demons that they could never be slain by a man. They considered a woman too weak to fear. Armed with this powerful boon, they created havoc in the Universe—destruction and hardship on the earth. They trampled the innocent, and they destroyed nature. Finally, they came face to face with Goddess Durga. Seeing the cruelty inflicted by Chanda and Munda on Her children; the face of the goddess Durga became dark with rage. At that moment, Goddess Kali jumped out from Her right eyebrow. The face and appearance of Goddess Kali were terrible. Merely Her sight caused fear and havoc in the army led by Chanda and Munda. Many of the demons in the army fell dead out of shock and fear. Those who did not die of fear ran away from the battlefield. She was wearing a long garland of severed heads, as well as a serpent necklace around Her neck. Those demons in the arms of Chanda and Munda, that tried to attack Goddess Kali, were killed by Her with ease. Goddess Kali then engaged in a ferocious fight with Chanda and Munda, who were powerless and clueless before the mighty Goddess Kali. Soon, Goddess Kali cut off the heads of the demons Chanda and Munda and presented them to the goddess Durga. Then came another demon called Raktabija (seeds of blood). Raktabija was wounded, but drops of blood falling on the ground created innumerable other demons (as per the boon granted to Raktabija). Goddess Kali devoured the duplicates of the demons into Her gaping mouth. This form, which drank the demon's blood, is also

DOI: 10.1201/9781003304814-8

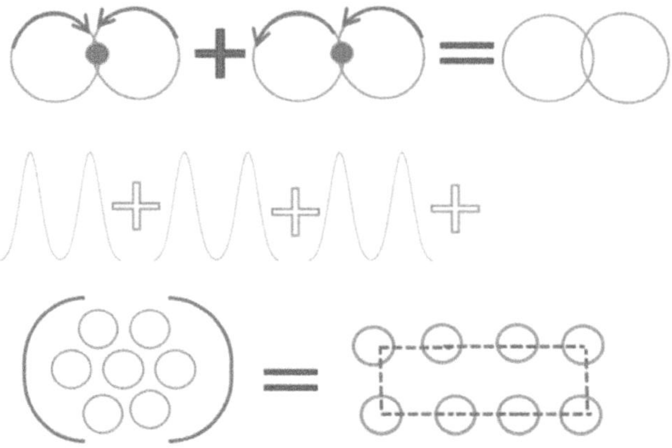

FIGURE 7.1
Vortex structure created in a spin-triplet superconducting material shows magnetic lines of field. Two sets of oppositely rotating circles around two red points can be added to create two intersecting not rotating circles. They are represented by the addition of two peaks. Several small circles contained in a cavity form an ordered state such as a rectangle decorated with circles.

called 'Rakteshwari'. Ultimately, Raktabija too was annihilated. I find this story can be compared with the particle interactions and phase transition, and I explain next.

Like light particles in a vacuum in materials, Boson particles having integer spins that can carry information; however, they can be prevented from doing so by other particles or defect centers that carry half-integer spins. In metals, the half-integer spin particles (spin of electrons) can follow over a long distance, but they can be stopped by a static spin center, which can form an integer spin (bosons). More electrons gather to overcome the spin of the defect center, and finally it happens. In a superconductor, a Cooper pair having a zero-spin state tries to overcome the resistance of the magnetic impurities (Tamas), which has polarity. Superconductivity can also be destroyed by an electric field or by charged particles having opposite polarities. This is represented by the spin-orbit (S-O) coupling that creates duality. A superconducting state tries to overcome the effect of S-O. In this story of Chandi (Kali), we see two pairs of demons (Shumbho–Nisumbho and Chanda–Munda). These two pairs remind me of a flower having four petals consisting of two types of dualities. This is a model of a *d*-wave superconductor. It forms a double degenerate state instead of a single degenerate state, as seen in the story of Goddess Durga. They can represent two magnets or two pairs of opposite spins. This model looks very similar to the spin-triplet state having two combinations of opposite spins, such as $|\uparrow\uparrow\rangle$ and $|\downarrow\downarrow\rangle$. We can have a third type of pairing $|\uparrow\downarrow\rangle + |\downarrow\uparrow\rangle$. This combination of polarity is stronger than a single pair of opposite spins. However, by destroying the spin-triplet states, many spins are produced. These spin centers try to oppose the flow of Cooper pairs. A strong force from the superconductors is required to destroy all these spin centers. With the help of their magnetic field, superconductors destroy all other magnets and establishe the flow of the Cooper pairs. The magnetic centers become a part of the superfluid. The attractive force between the electron-hole pairs is the most crucial to establishing the strength of the superconductors. This attractive force originates from the exchange of phonons or heat from the lattice. We consider heat as Shakti, the primordial form of energy.

In Chapter 3, we said that the Universe is created from a fluid body. Here we apply the concept of the quantum fluid consisting of spin-like liquid bound to the charge or orbits. These fixed points can form a vortex, which in the end is an absolute static object surrounded by spin dynamics. This model is described in Sankhya philosophy. The singularity is the most massive part (Tamas), whereas the rotating spin liquid can be compared to Rajas. The emission of the magnetic field of light from the vortices can be compared to the Satwa guna of the light.

The shell model described in Sankhya philosophy, having different energy levels, should be associated with the emission or absorption of electromagnetic radiation (or light) of different frequencies. This active part is dynamic in nature and cannot be manifested but maintained at the state of equilibrium. This maintains different levels of energy, the lowest one being Tamas, which works as a deep potential or attractive one. The second one is repulsive, corresponds to a higher state of energy, and is called 'Rajas'. The highest state of energy is Sattwa, which maintains the equilibrium of strong attractive and repulsive forces. However, the state of equilibrium can be destabilized, and again it starts to regain the equilibrium. This process creates the emission of waves as light. This reminds us of the creation of stars from a cloud of particles associated with the emission of radiation. In condensed matter physics, the creation process can be described on the microscopic scale, as described in this chapter. Mass is created from energy. Forces are carried away with the help of elementary particles to form the particles associated with matter. I find this conversion has a close link to the creation process as interactions between the Purusha and Prakriti (as two different particles) are described in Sankhya philosophy. Let me elaborate on the particles followed by the condensate.

Bosons and Fermions

There are possibly two classes of particles in the Universe:fermions and bosons. A fermion is any particle that has an odd half-integer (like 1/2, ¾, 2/3, and so forth) spin. Quarks and leptons, as well as most composite particles like protons and neutrons, are fermions. The consequence of an odd half-integer spin is that the fermions obey Pauli's exclusion

FIGURE 7.2
Cooper pairs are associated with a superconducting state. The dancing of Cooper pairs is strongly correlated. This interaction can be compared with dancing of couples. It seems their movements are synchronized.

principle and therefore cannot coexist in the same state at the same location at the same time. On the other hand, bosons are those subatomic particles whose spin quantum number has an integer value (0, 1, 2, ...). All the force carrier particles are bosons, as are those composite particles with an even number of fermion particles (like mesons). The scheme of Quantum field theory is that fermions interact by exchanging bosons. Fermions are solitary. Only one fermion may occupy any quantum state. The fermionic solitariness of electrons is responsible for the structure of molecular matter (in fact, for all 'structure' in the Universe). The degeneracy pressure that stabilizes white dwarf and neutron stars is a result of fermions resisting further compression toward each other. Fermions obey Fermi–Dirac statistics.

Bosons may occupy the same quantum state as other bosons, for example, in the case of laser light, which is formed of coherent, overlapping photons. The more bosons there are in a state, the more likely it is that another boson will join that state (Bose condensation). Fermions are usually associated with matter, while bosons are the force carriers. The electrons belong to the class of elementary particles called 'leptons'. Leptons and quarks, together, constitute the class called 'fermions'. According to the Standard Model, all mass consists of fermions. Whether the fermions combine to form a table, a star, a human body, or a flower or do not combine at all depends on the elementary forces—the electromagnetic, the gravitational, the weak, and the strong forces. All force is mediated by the exchange of gauges or bosons. The electromagnetic force is mediated by the exchange of photons, the strong force by the exchange of gluons, and the weak force by the exchange of W and Z bosons.

The exchange interactions among particles are the key to the creation process (we shall elaborate on this through the uncertainty principle). A quantum vortex will be formed in the Bose–Einstein condensate based on this. Some bosons are elementary particles and occupy a special role in particle physics, unlike fermions, which are sometimes described as the constituents of ordinary matter. Some elementary bosons act as force carriers, which give rise to forces between other particles, while one (the Higgs boson) gives rise to the phenomenon of mass. Other bosons, such as mesons, are composite particles made up of smaller constituents. Superfluidity arises outside the particle physics realm because composite bosons (Bose particles), such as low-temperature helium-4 atoms, follow Bose–Einstein statistics; similarly, superconductivity arises because some quasiparticles, such as Cooper pairs, behave in the same way.

Pauli's Exclusion Principle

Pauli's exclusion principle is the quantum mechanical principle that two or more identical fermions (particles with half-integer spin) cannot simultaneously occupy the same quantum state within a quantum system. In other words, (1) no more than two electrons can occupy the same orbital, and (2) two electrons in the same orbital must have opposite spins. Pauli's exclusion principle plays an important role in the understanding the electronic structure of molecules as it does in the case of atoms. The effect of Pauli's exclusion principle is to limit the amount of electronic charge density that can be placed at any one point in space. This principle applies to identical particles with half-integral spin, i.e., $S = ½, 3/2, 5/2$. In other words, each electron should have its singlet state or unique state.

Particles with an integer spin or bosons are not subject to Pauli's exclusion principle: Any number of identical bosons can occupy the same quantum state, as with, for instance, photons produced by a laser or atoms in a Bose–Einstein condensate. A more rigorous statement is that, concerning the exchange of two identical particles, the total (many-particle) wave function is antisymmetric for fermions and symmetric for bosons. This means that if the space and spin coordinates of two identical particles are interchanged, then the total wave function changes its sign for fermions and does not change it for bosons. If two fermions were in the same state (for example, the same orbital with the same spin in the same atom), interchanging them would change nothing, and the total wave function would be unchanged. The only way the total wave function can both change sign as required for fermions and remain unchanged is that this function must be zero everywhere, which means that the state cannot exist. This reasoning does not apply to bosons because the sign does not change.

Many mechanical, electrical, magnetic, optical, and chemical properties of solids are the direct consequence of Pauli's exclusion principle. The stability of each electron state in an atom is described by the quantum theory of the atom, which shows that a close approach of an electron to the nucleus necessarily increases the electron's kinetic energy, an application of the uncertainty principle of Heisenberg. However, the stability of large systems with many electrons and many nucleons is a different question and requires Pauli's exclusion principle, which is responsible for the fact that ordinary bulk matter is stable and occupies volume. Also, astronomy provides a spectacular demonstration of the effect of Pauli's principle in the form of white dwarfs and neutron stars. In both bodies, the atomic structure is disrupted by extreme pressure, but the stars are held in hydrostatic equilibrium by degenerating pressure, also known as Fermi pressure. This principle would apply to the stable structure of a quantum vortex or a black hole, discussed in Chapter 9.

Comment: A pair of spin half particles produces a zero-spin state (or a degenerate state), which follows the dualistic model, as discussed. This pair forms a polarity (a dipole) that cannot accept the third spin. If we combine more spins, they form a spin-one (spin-triplet) state that loses duality (polarity). Therefore, many spin-neutral particles can be accommodated in a state (a doubly degenerate state).

Bose–Einstein Condensate (BEC)

In condensed matter physics, a Bose–Einstein condensate (BEC) is a state of matter (also called the 'fifth state' of matter), which is typically formed when a gas of bosons at low densities is cooled to temperatures very close to absolute zero (−273.15 degrees centigrade, −459.67 degrees Fahrenheit). Under such conditions, a large fraction of bosons occupy the lowest quantum state, at which point microscopic quantum mechanical phenomena, particularly wave function interference, become apparent macroscopically. A BEC is formed by cooling a gas of extremely low density (about 1/100,000 part of the density of the normal air) to ultralow temperatures. The state was first predicted by Albert Einstein in 1925, following and crediting a pioneering paper by S. N. Bose on the new field, now known as quantum statistics. The result of this effort is the concept of a Bose gas, governed by Bose–Einstein statistics, which describes the statistical distribution of identical particles with integer spin, now called bosons. Bosons, particles that include the photon as well as atoms such as helium-4, are allowed to share a quantum state. Einstein proposed that cooling bosonic atoms to a very low temperature would cause them to fall (or condense) into the lowest accessible quantum state, resulting in a new form of matter. In 1938, fritz London proposed the BEC as a mechanism for superfluidity in helium-4 and superconductivity. Vortices in Bose–Einstein condensates are also currently the subject of analog gravity research, studying the possibility of modeling black holes and their related phenomena in such environments in the laboratory. Studies of vortices in nonuniform Bose–Einstein condensates, as well as excitations of these systems by the application of moving repulsive or attractive obstacles, have also been undertaken. Within this context, the conditions for order and chaos in the dynamics of a trapped Bose–Einstein condensate have been explored.

BEC is the lowest-energy or the ground state of the bosonic system and has poor stability and can be easily distributed by perturbation. Vortices can be formed in the BEC, which can be described as the battle between the gods (bosons) and demons (fermions), having an integer and half-integer spin, respectively. The BEC looks like the fluid that remains still (state). Sankhya philosophy says Purusha remains (absolutely) static; however, it can be disturbed, leading to a state of nonequilibrium. Sankhya philosophy has also elaborated the interactions between static space and time (dynamics), which gives the distortion of space–time or uncertainty. Duality is always associated with nature and can be described by the uncertainty principle.

Uncertainty Principle

In quantum mechanics, the uncertainty principle (also known as Heisenberg's uncertainty principle) is any of a variety of mathematical inequalities, asserting a fundamental limit to the accuracy with which the values for certain pairs of physical quantities of a particle, such as position and momentum, can be predicted from initial conditions. Such variable pairs are known as complementary variables or canonically conjugate variables, and, depending on interpretation, the uncertainty principle limits to what extent such conjugate properties maintain their approximate meaning, as the mathematical framework of quantum physics does not support the notion of simultaneously well-defined conjugate properties expressed by a single value. The uncertainty principle implies that it is in general not possible to predict the value of a quantity with arbitrary certainty, even if all initial conditions are specialized. Heisenberg's uncertainty principle states that there is inherent uncertainty in the act of measuring a variable of a particle. Commonly applied to the position and momentum of a particle, the principle states that the more precisely the position is known, the more uncertain the momentum is and vice versa. Heisenberg's uncertainty principle is a key principle of quantum mechanics. Very roughly, it states that if we know everything about where a particle is located, the uncertainty of position is small; if we know nothing about its momentum, the uncertainty of momentum is large and vice versa.

Introduced first in 1927, by the German Physicist Werner Heisenberg, the uncertainty principle states that the more precisely the position of some particle is determined, the less precisely its momentum can be predicted from initial conditions and vice versa. Thus the product of the standard deviation of position and the standard deviation of momentum are greater than and equal to half of the value of the reduced Planck constant. Historically, the Uncertainty principle has been confused with a related effect in physics, called the observer effect, which notes that measurements of certain systems cannot be made without affecting the system, that is, without changing something in the system. Heisenberg utilized such an observer effect at the quantum level as a physical explanation of quantum uncertainty. It has since become clearer, however, that the uncertainty principle is inherent in the properties of all wave-like systems and that it arises in quantum mechanics, simply due to the matter–wave nature of all quantum objects. Thus the uncertainty principle states a fundamental property of quantum systems and is not a statement about the observational success of current technology. It must be emphasized that measurement does not mean only a process in which a physicist-observer takes part but rather any interaction between classical and quantum objects regardless of observer.

Since the uncertainty principle is a very basic result in quantum mechanics, typical experiments in quantum mechanics routinely observe its aspects. Certain experiments, however, may deliberately test a particular form of the uncertainty principle as part of their main research program. These include tests of number–phase (ΔN-$\Delta \Phi$) uncertainty relations in superconducting or in systems of quantum optics. Applications dependent on the uncertainty principle for their operation include extremely low-noise technology that is required in gravitational wave interferometers.

Kondo Scattering

The word 'Kondo' means battle in Swahili. In correlated electron physics, the Kondo effect is a standard model that describes the scattering of conduction electrons due to the presence of magnetic impurities, resulting in a characteristic change in the electrical resistance with temperature. For normal metals, the scattering of conduction electrons is reduced at low temperatures, resulting from a reduction in electron-phonon scattering; consequently, electrons propagate with less backscattering. It was first described by Jun Kondo using the perturbation theory for dilute magnetic alloys where he hypothesized that the scattering rate of the magnetic impurity should diverge as the temperature approaches absolute zero. This is caused by the conduction electrons that tend to align their spins opposite to that of a nearby impurity. At low temperatures, the electron propagation is deviated by a magnetic impurity that consequently flips its spin and deflects its pathway. The delocalized electrons pair with the local electron to form a state of total spin zero: a singlet that is the cause of the dramatic change in the resistance as the temperature decreases in contrast to a decrease in resistance in pure metals at low temperatures. Hence, in essence, a battle ensues when a magnetic impurity is placed in nonmagnetic metal. The Kondo effect does not depend on the properties of the magnetic impurity and the nonmagnetic host, but it is also accepted by both the size of the sample and the nonmagnetic random scattering events. At low temperatures and low energies, the interaction between the itinerant electrons and the localized magnetic impurity is nonperturbatively strong, a condition known as 'asymptotic freedom'.

There are two approaches to the Anderson model: the atomic picture that considers the tune hybridization picture and the adiabatic picture that considers the tune interaction strength. The discrepancy between the two pictures is resolved by considering the tunneling of local moments between spin up and spin down.

Kondo's phenomenological model was shown to emerge from Anderson's more microscopic model: The spin of delocalized electrons is opposite to that of the local electrons due to the tunneling of electrons on and off the local site. The Kondo effect has been observed in heavy fermions and intermetallic insulators that contain rare earth elements. The Kondo effect need not involve the screening of a localized spin but rather the screening of any localized degeneracy as the effect has also been observed when an even number of electrons occupy a quantum dot, at a singlet-triplet degeneracy. The quantum dot with at

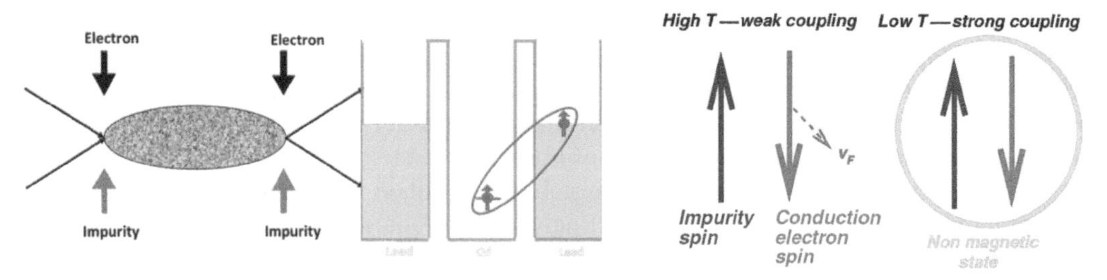

FIGURE 7.3
Demonstration of electron impurity-scattering via spin exchange. Demonstration of the Kondo effect in a zero-dimensional quantum dot system connected to electrodes, which arises when an unpaired spin from the quantum dot is paired to an itinerant electron to form a Kondo system. (left) Electrons and impurities represented by down and up arrows compete around a region or a shielded area. (middle) In a triple well structure they interact across the barrier and for a pair. The coupling between the spins becomes stronger at low temperatures.

least one unpaired electron is analogous to the traditional Kondo system of the magnetic impurity in metal. It behaves as a magnetic impurity, and, when the dot is coupled to a metallic conduction band, the itinerant electrons can scatter off the dot.

Quantum Vortex

The creation model can be elucidated as the rise of Purusha from a cosmic bath by overcoming duality, which finally stabilizes the whole system. It is said that nothing is created or destroyed but is a conformal change of space. However, space expansion is followed by a contraction, and the matter goes into the state of a small dense object: a black hole. It is a holographic world in a black hole. The core or singularity is Satwa, the rotation in the hole is Rajas, and the outer covering like the event horizon is Tamas. A black hole consists of a great zero, Mahasunya at the deepest point surrounded by an outer layer, i.e., Sunya. Within a black hole, the world is multidimensional, the effect of folding a flat space again and again. Space expands by division. This is an illusion that creates multidimensionality. The holographic world can be formed by rotation. Two identical images of different colors or rotated slightly give an extra dimension. So division plus rotation can give higher dimensions like a tunnel or funnel or a coiled serpent. This is the structure of a vortex where the core is undistributed. This is like the creation of fire by breaking up molecules into smaller pieces and/or rotating two surfaces in opposite directions.

In condensed matter physics, the great 'illusion' or flip may be described as an exchange interaction through spin or phonons so that a Cooper pair can be formed. It is like a small vibration or heat energy or even magnetic flux (a phase) that binds the fermions together. So the story can be rewritten in particle physics as the conversion of bosons (spin 0/1) to fermions (spin 1/2). In one state, two electrons of opposite spins can be present, like (↑↓). The half spins have equal mass but opposite polarities. They can have as many as possible exchange interactions and bosons (zero or integer spins) in one level, just as many stars in the sky are allowed but one sun (Purusha) and Shakti (its heat or other characteristics). From a spin-singlet (↑↓) state, a spin-triplet state can be created where two electrons of

FIGURE 7.4
Collections of microscopic fire form a macroscopic state of fire that can spread over a wide region.

equal polarity can exist; however, it is compensated by two other spins of equal and opposite polarity, like |↑↑> and |↓↓>. You can have a third type of pairing |↑↓> + |↓↑>, which is similar to the subject. Spin $m = 0$ means a full flux, which gives a massless particle like a photon. For singlet $s = 0$, where one up spin and one down spin attract each other with the help of Shakti, the strong attractive force of photon. In triplet state $s = 1$, full flux is also formed with a (↑↑+↓↓) configuration. To create massive particles or fermions, we need a half-flux. Shakti converts bosons to Fermions by changing a full flux to a half-flux.

Vishnu lies in the ground state (bosonic) (E_P), which is static and looks like a bosonic ocean, a superfluid or condensed state like a water bed when time has not been started. Bosons can be massless particles like photons. Brahma is created from Vishnu, which is some massive particle like fermions (not Cooper pairs) in superconductivity. The current of these particles is somehow blocked because of noise created through interference.

Goddess Kali is Sakti living within Vishnu and is released from the bosonic bath. She is some kind of sound energy that binds particles together. She creates glue between two particles or atoms. In physics, this is described as an exchange interaction through spin or phonons so that a Cooper pair can be formed. It is like a small vibration or heat energy or even magnetic flux (a phase) that binds the fermions together. The demons formed as Madhu-Kaitava are also bosonic particles that are part of Vishnu since they are derived from the ears of Vishnu. But they can destroy the tunnel current of Brahma through interference. So the story can be rewritten in particle physics as the conversion of bosons (spin 0/1) to fermions (spin 1/2). Spin $m = 0$ means a full flux, which gives a massless particle like a photon. For singlet $s = 0$, one up and one down spin attract each other with the help of the goddess, the strong attractive force of photon. In triplet state $s = 1$, full flux is also formed with a (↑↑+↓↓) configuration. To create massive particles or fermions, we need a half-flux. Goddess Kali converts boson to Fermions by changing a full flux to a half-flux. This is connected to pure sound. Full flux is like Karma; someone puts the electron in an orbit that will rotate on and on. The radiation emitted from the rotation will not return and will be liberated. Here two fermions have no attraction or repulsion since they

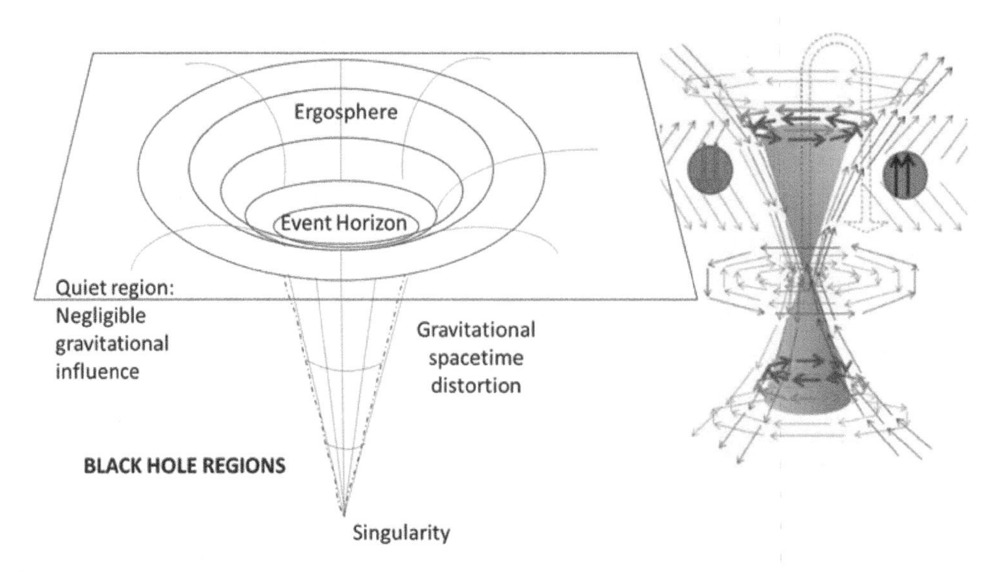

FIGURE 7.5
A black hole is represented by an infinitely deep funnel. It starts from a planar undisturbed region and ends to a point which is called singularity. This can be compared with the vortex structure created in a spin-triplet superconducting material. Hence a holographic universe can be imagined.

like to work without any benefit. This is the condensed state, which can be free from disorder. When time has not started, the situation was like a condensed state. No work has been done. Inertia stops working or stops moving objects. The noise stops the creation of pure sound or bosons. Goddess Kali delivers kinetic or heat energy to Vishnu, producing heat. Vishnu brings an extra phase by Sudarshan Chakra ($\Delta\phi$) rotation, which kills the demons, creating interference, and converts full flux to a half-flux; therefore, massive particles are produced. Sakti or Goddess Sakti has many forms. It is a force that is created by breaking the space like beta decay in the weak nuclear force. Besides breaking the body of Vishnu like a nuclear force, this nature is also seen in Uma, the wife of Shiva, whose body was divided into many parts by Vishnu. There, Sati or Uma in the form of a corpse was a symbol of space. Shiva was like heat energy that was excited and rotating the space like the Universe. Vishnu created many parts like stars and planets from the state, which de-excited Shiva finally.

Shakti is also like a strong force that binds particles together—the ultimate glue. Shakti is the force that creates rotation; Vishnu used his weapon, the Sudarshan Chakra, to control the Universe and a phase. This is like a magnetic field. Sakti is like gravitation as it attracts and rotates the world. So it has the properties of all fundamental forces. But her best property is Great Illusion. So it is nonlocal in superconductors where duality works. $\Delta\Phi\Delta N \geq 1$ is the property of Cooper pairs. $\Delta\Phi = 0$ is a supercurrent (Kali), or $\Delta N = 0$ is a super insulator (Shiva). Brahma is the manifestation of a Cooper pair that is tunneling through the insulator (Vishnu), which is acting as the preservation of energy in the form of a bosonic bath. This is possible through the supercurrent of Goddess Kali, which works against the normal current. We see these opposite forces everywhere, and the goddess must kill any other opposing processes. Additional effects, such as Kondo resonance, due to magnetic impurities can be explained with this model. All these transition features can be explained by vortex–antivortex interactions. Vortex means the binding of magnetic flux in the superconductors. It has a rotating component and strong attractive force, and it breaks up condensates through the addition of a magnetic center that produces the fields. I consider vortex behavior as the manifestation of Goddess Kali, which tunnels through the superconductors. At low temperatures, the superconductor goes to the perfect condensate without the presence of a vortex. This is a state of inactive Vishnu in deep sleep, which is Tamas or Satwa. A vortex is manifested if the temperature is slightly increased, and the resistance is also increased below the onset of the superconductivity transition. There will be vortex–antivortex interaction. Vortex–antivortex fusion produces attraction, which is a signature of Rajas. Above the transition temperature, the vortex and antivortex repel each other, which is like Tamas or Satwas since resistance is manifested. Anyway, a vortex arises from rotation thst counteracts the exchange interaction. It is like an elementary particle.

Spin-Flip Interactions

Exchange interactions are the best example to prove Vishnumaya since it creates a phase or rotation to the system with strong interactions between spins. Depending on the pairing potential, a singlet or triplet state or ferromagnetic order can be seen. This can also explain the formation of energy bands. Madhu-Kaitava is like doping that disturbs the system in a steady state, brings disorder, and tries to stop the formation of condensate (Cooper pair) from the fermionic bath, which is like the eternal bed of Vishnu. The Kondo impurity creates an impure band surrounded by electrons. The Kondo impurity fights Vishnu (the sea of electrons when excited). This is like an exchange interaction (like the fight between Vishnu and Madhu-Kaitava). Initially, Kondo is stronger than the electrons, but the electrons absorb energy from the lattice or Vishnu. So heat is released from Vishnu. Heat represents Kali, who brings the exchange in the fermionic bath (Vishnu) and goes to the excited state. Kondo impurity, spin, and electron spin try to flip one another. These spins in opposite directions due to flip create an antiferromagnetic state, which is due to the rotation of spin (Goddess Kali), and Brahma is born as an AFM state. Another flip of AFM gives superconductivity; this flipping is Maya for one center and Mahamaya for the whole system. She occupies the space between an electron and a hole. This spin or flip is equivalent to the self-spin of Goddess Kali. Two spins in a ferromagnetic material are localized or locked, forming an insulating or inactive state (like Vishnu) but from a spin-flip (spontaneously) antiferromagnetic state. Further spin flips yield superconducting, delocalized current. Vishnu is dreaming but cannot create matter or Brahma due to the

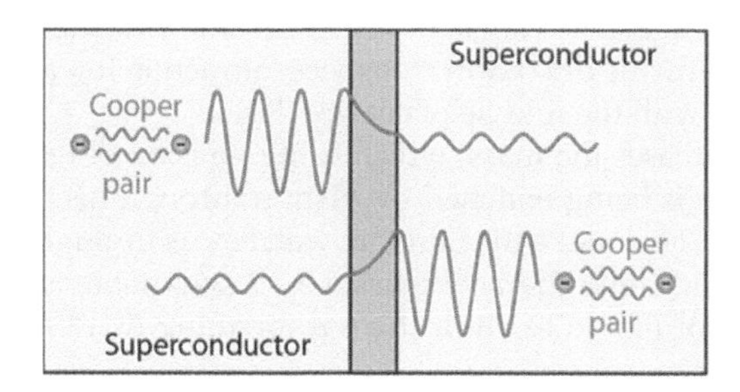

FIGURE 7.6
Two electrons coupled by a wave form a pair, called a Cooper pair in a superconductor. The wavefunction that describes a Cooper pair of electrons in a superconductor is an exponential like the free particle wavefunction. A wavy line represents the wavefunction of the pair which penetrates a barrier between two superconducting materials although the amplitude of the tunnel current diminishes.

noise of the demons. Vishnu becomes angry, and heat is released from his body (Kali), which kills noise.

The Holographic Universe

Vishnu rises, overcoming duality (dream) and stabilizing the systems. Cooling produces matter like iron from ore. It is said that nothing is created or destroyed but is a conformal change of space. Kali is produced from Durga, Vishnu from Mahavishnu (or the absolute Vishnu), and Shiva from Sadashiva (or the absolute Shiva). Vishnu's dream and rise add an extra dimension to the flat space and makes it curved. The dream is our wish that we cannot create in reality can be formed. A massless boson is a dream that gives enormous pleasure, and fermions are a reality. Dreams can be multidimensional. During our sleep, heat (boson) is released from the body, cooling the body and enabling more dreams to come into the mind like bubbles. The body expands, followed by contraction, and matter goes into the state of a small dense object, a black hole. Good dreams need no disturbance, and no interference can destroy disturbance. It is a holographic world in a black hole. The core or singularity is Satwa, the rotation in the hole is Rajas, and the outer covering like the event horizon is Tamas. A black hole consists of a great zero, Mahasunya at the deepest point, surrounded by an outer layer, i.e., Sunya. Within a black hole, the world is multidimensional, which can be produced by folding a flat space again and again. Space expands by division. This is an illusion that creates multidimensionality. The holographic world can be formed by rotation. Two identical images of different colors or rotated slightly give an extra dimension. So division plus rotation can give higher dimensions like a tunnel or funnel or a coiled serpent. From a coiled snake that is binding the space (like Vishnu sleeping on a coiled snake), the head of the snake goes to infinity or singularity. This clearly explains the uplifting of Brahma from Vishnu from his dream through Kali who rotates and divides the space. This is the structure of a vortex where the core is undistributed. This is like the creation of fire by breaking up molecules into smaller pieces and/or rotating two surfaces in opposite directions.

Maya is like holography, which means two particles are rotating in opposite directions at the same time. This is a duality that can create a three- or multidimensional space through the possible emission or absorption of energy like heat emission, which flows back to the system. This backflow is precisely the nature of Kali who maintains a feedback loop. The surface of the event horizon is flat or two-dimensional, which becomes three-dimensional at the black hole. As space merges with the black hole, some energy comes out because of feedback (backflow). This is like Tamas or Great Illusion (Mahamaya). Since the dimension increase is assisted by the breakdown of space, attraction toward the center increases. Finally, it breaks down with the release of energy (Fire).

Folding the space increases the mass, which is the division of itself or a zero (vacuum). This folding or division is being watched by Vishnu through his dream, which goes to a singular point: Brahma. This is like the Sun that watches us throughout time in association with the night (Kali) and three characteristics, i.e., dawn, noon, and dusk. Everybody is watching the progress of life on earth. If there is anything that blocks the vision, then it will be killed.

Focusing our vision on a point is the most important thing. Two eyes (even three) focus at a point, creating the third dimension. Vishnu watches the progress of Brahma (creator of life) like the Sun from the center of the Universe. Shiva watches Kali (nature or Prakriti) all the time. God watches His creation after creation. Finding a singularity

FIGURE 7.7
One black four handed goddess is born from the third eye of another ten handed goddess who fights demons and kills them. She is holding different weapons in her four hands. This process can be compared with the conversion of spin 1 particles into spin half particles to counteract the magnetic impurities.

is not the end of the story. Once all energy is concentrated in some area, it radiates and goes to another singularity. After passing through the singularity, the rays can diverge. From one concentrated point, there can be a (sudden) collapse. This is said to be the rebirth of Brahma. Also, Kali emits fire through her tongue, which can be pointed to even the interference of the teeth in her mouth. This is called 'resonance', and sound is created from the resonance. After reaching singularity, energy flows back, which can be imagined from the shape of the flame in a candle or similar places. Rebirth is shown as backflow or water coming out from the head of Shiva or the serpent on the head of Vishnu. This is also matched with the emission of rays from black holes. All these birth processes need the contraction of the mind. Focusing all energy or heat means Anger or Ego, which is the cause of creation. Vishnu was dreaming, and he did not want to be disturbed before creating Brahma. He became angry with the disturbance of Madhu-Kaitava, and he released heat (the goddess) from his body, which assisted the creation of sound through resonance: Brahma. Goddess Kali heats the process of creation, which attains equilibrium.

At the initial state, a layer of gaseous material was stagnant. There was some heat energy in the Universe, but it was distributed so the temperature was not raised, and there were no transformations of states. Due to a disorder in the gaseous system, the symmetry of the arrangement was broken, creating self-spin-like bubbles in the water. The bubbles create space inside, which needs to be filled up but never collapses. One imperfection of voids in the sea of the mind asks, "Give me energy! I am lacking energy."

Adding energy from all sides creates this bubble that is associated with self-spin since it is not stable. This produces a magnetic field (B). The magnetic field moves out of the sea of the mind. Two opposite rotations create Maya. Ultimately, the whole system becomes a sphere consisting of all bubbles. All fields concentrate at a point and create a very bright field. (Refer to Figure 7.6.)

This is like concentric circles or one circle containing many small circles. A garland contains several gemstones (or flowers). One big zero contains many zeroes. Since the fields are focused, a new zero is created, i.e., the sun. The sun is a part of the Brahmanda and is associated with thousands of heads (bubbles) and of hands/legs (field lines or solar rays). These bubbles are called 'Atman', which is a part of Absolute Atman. When this Brahmanda dissociates, it can first create a dual like the shifting of a ball slightly from its position. Then each of the displaced images split into two. Following this creation of the Brahmanda, all bubbles can be split into two and then into four parts.

This large sphere of Brahmanda cannot go everywhere. So it splits into parts and tunnels through part by part. Thus the body of Vishnu is split. This splitting creates two opposite circular paths. One is Mahamaya, i.e., from bubbles to Brahmanda. Maya, or the dream of Vishnu, is overcome by Mahamaya, which compensates two Maya, i.e., creating a large space when Vishnu rises. Maya is the phase of the moon. Mahamaya is the phase of the sun that suppresses the lunar phase. So two zeroes (two Maya) are connected to form Mahamaya. This is like linking up one vortex and one antivortex, which adds an extra dimension. From a flat circle, we get a fountain-like space. This creates equilibrium, overcoming all disturbances. Goddess Kali is self-spin and connects all zeroes (spins), moves all zeroes together, puts them together in the sky, and Shiva dissociates them. This is just like a rain cycle. The body of Vishnu is the bed of water. It cannot go up to the sky itself and remains static. The sun (beam) strikes the water and creates heat in the body of water. Heat dissociates the water body of Vishnu (Narayan) into small water droplets (like zeroes). They can fly up to the sky (tunneling in parts through the air) since Goddess Kali (heat) breaks up the body of Vishnu. However, the cloud may not transform into rain. Polarity in the cloud has to be created, which breaks up the cloud with sound. Raindrops come back to earth. The flow of water molecules upward and falling are controlled by Goddess Kali as the current. The body of the cloud covers the face of the Sun, which is destroyed by Vishnu. The cloud is converted into rain and the Sun appears to us as Brahma. The condensation of small molecules of water into a large one is the Maya-to-Mahamaya transition. The path of the water molecule to the sky is the rise of Vishnu. The body of a black cloud is Madhu-Kaitava, two demons that are created from Vishnu and that create interference from the sunbeams, so the Sun is not properly visible. What Indra could not create (a proper cycle) Vishnu did with the help of Mahamaya, a consort of Vishnu. The rain cycle can be described better with one upward and one downward direction as if one column is produced from one vortex and another column is produced from an antivortex. Thus we have two cycles orthogonal to the third cycle like a fountain. This is like a third eye. Two eyes correspond to a vortex and an antivortex, and the third one binds them together. This is like a triplet structure with three characteristics produced from duality or a doublet structure. This model explains the Rebirth of Brahma. Goddess Kali is the energy gap or makes the gap. It is mostly the energy gap that creates orbits around the center, i.e., the Sun or nucleons. In energy bands, holes and electrons are also called 'filled' or 'unfilled' bands. Since there is a gap, in-between light emission (or absorption) can take place. Vishnu on the filled band looks for the unfilled band with the desire to occupy the unfilled band. It is the wish of Vishnu that splits energy into filled and unfilled states and the gap. Vishnu consorts with the gap and unfilled states. So electrons are emitted across the gap from the filled band. This transition makes the emission of light, the rise of Brahma. The rebirth of Brahma can be explained by the transition of electrons at different levels in the conduction band. Finally, it reaches a continuum state known as a 'halo' or the 'ultimate shield' of the band. Goddess Kali or Atman occupies the space between e^- and h and forms a Cooper pair. By exchange interactions between these two particles, ferromagnetic or antiferromagnetic states are formed.

In the previous discussion of Sankhya philosophy, we compared it to the quantum vortex model. One of the best applications of Sankhya philosophy is spin-orbit coupling, which combines electric (orbit-like) and magnetic (spin-like) fields. The SO coupling is essential for creating a vortex field that consists of the static axis surrounded by the spins. Two vortices give a spin-triplet structure, discussed thoroughly in Chapter 8. Let me elaborate on the concept of spin-orbit coupling first.

Spin-Orbit Coupling

Spin-orbit coupling is the interaction between the electron's spin and its orbital motion around the nucleus. In quantum physics, the spin-orbit interaction (also called spin-orbit effect or spin-orbit coupling [SOC]) is a relativistic interaction of a particle's spin with its motion inside a potential. This spin-orbit coupling is one of the relativistic effects, and it occurs whenever a particle with nonzero spin moves around a region with a finite electric field. A semiclassical model is described as the origin of the spin-orbit interaction in a simple system such as a hydrogen atom. The interaction energy is due to the coupling of the induced electric dipole with the electric field of the nucleus. In atomic nuclei, the spin-orbit interaction is much stronger than for atomic electrons and is incorporated directly into the nuclear shell model. (Refer to Figure 7.5.)

When an electron moves in the finite electric field of the nucleus, the spin-orbit coupling causes a shift in the atomic energy levels of the electrons due to the electromagnetic interaction between the spin of the electron and the electric field. In the rest frame of the electron, there exists a magnetic field created by the interaction of the angular momentum of the electron and the electric field of the nucleus. The electric field in this case has various physical origins such as the electric field of the atomic nucleus or the band structure of a solid. The spin-orbit coupling increases with the atomic number of the atom. Spin-orbit coupling can be regarded as a form of effective magnetic field seen by the spin of electrons in the rest frame. Based on the notion of the effective magnetic field, it will be straightforward to conceive that spin-orbit coupling can be a natural, nonmagnetic means of generating spin-polarized electron current. In Chapter 8, we shall see how SOC influences the properties of a superconductor and turns it into a spin–triplet state.

Summary

In this chapter, we discuss more examples of the dualistic view in terms of elementary particles, i.e., bosons and the fermions, which are associated with integer values and half-integer values of the spins. In condensed matter, there is competition between the two types of particles that deliver a phase transition and all other exotic quantum features, such as spin-triplet superconductivity. Understanding the competition will present a clear picture of quantum mechanics, particularly the uncertainty principle. The competition can be described by spin-orbit coupling, which describes the vortex model in a very scientific manner. As a result, we see spin-triplet superconductivity. In Chapter 8, we shall explain the basic concept of quantum mechanics and quantum tunneling.

8

Creation through Interactions at the Microscale—Quantum Mechanics

- **Quantum mechanics:** Matter–wave
- **Quantum tunneling**
- **Superconductivity**
- **Josephson tunneling**
- **Qubits (three cycles: σ_x, σ_y, σ_z)**

In quantum mechanics, since the particle size is very small, the transformation rate is very fast, reaching the speed of light. Prakriti interacts with Purusha at a great speed, a wave that changes to indistinguishable particles. The transformation of Satwa, Rajas, and Tamas also works at a very high speed so that they become inseparable. All elementary particles can be described as a unit of three, like quarks or electron-proton-neutron. In the atom (or in a nucleus), the elementary particles are arranged in different layers separated by spaces that can develop a shell structure. The energy from one shell to another can be transformed by overcoming the vacuum barrier. Here, the barrier plays the most important role for the particles to tunnel from one side to another. The barrier remains static (as an observer or neutral), which is described as the 'Purusha' in Sankhya philosophy. The tunnel current is Prakriti, which appears as the particles present on both sides of the barrier. This looks like the barrier is surrounded by particles. Black body radiation is the direct consequence of Sankhya philosophy, which originates in quantum theory as a combination of wave–particle duality. Hence, most importantly, quantum and resonant tunneling will be discussed in this chapter through Sankhya philosophy. Therefore, I shall introduce the basic principles of quantum mechanics followed by the quantum tunneling phenomenon.

Quantum Mechanics

Quantum mechanics is the fundamental theory in physics that describes the physical properties of nature at the scale of atoms and subatomic particles. It is the foundation of all quantum physics including quantum chemistry, quantum field theory, quantum technology, and quantum information science. Classical mechanics, the description of physics that existed before the theory of relativity and quantum mechanics, describes many

DOI: 10.1201/9781003304814-9

aspects of nature at an ordinary (macroscopic) scale, while quantum mechanics explains the aspects of nature at small (atomic and subatomic) scales, for which classical mechanics is insufficient. Most theories in classical physics can be derived from quantum mechanics as an approximation, valid at a large (macroscopic) scale. Formerly, scientists developed a consistent theory of the atom that explained its fundamental structure and its interactions. In 1926, physicists developed the laws of quantum mechanics, also known as the 'laws of wave mechanics'. The laws of quantum mechanics explain the atomic and subatomic phenomena. Quantum mechanics is the physics that explains how everything works: the best description we have of the nature of the particles that make up matter and the forces with which they interact. It characterizes the simple things, such as how the position or momentum of a single particle or group of few particles change over time.

Matter Waves

At the end of the 19th century, light was said to consist of waves of electromagnetic fields that propagated according to Maxwell's equations, while matter was thought to consist of localized particles. In 1900, this division was exposed to doubt; when investigating the theory of black-body radiation, Max Planck proposed that light is emitted in discrete quanta of energy. His notion was thoroughly challenged in the year 1905. Extending Planck's investigation in several ways, including its connection with the photoelectric effect, Albert Einstein proposed that light is also propagated and absorbed in quanta, now called photons. This quantum would have energy given by the Planck–Einstein relation.

Matter waves are a central part of the theory of quantum mechanics, being an example of wave-particle duality. All matter exhibits wave-like behavior. For example, a beam of electrons can be diffracted just like a beam of light or a water wave. In most cases, however, the wavelength is too small to have a practical impact on day-to-day activities. The concept that matter behaves like a wave was proposed by French physicist Louis de Broglie in 1924. Matter waves are referred to as 'De Broglie waves'. The De Broglie wavelength is the wavelength associated with a massive particle (i.e., a particle with mass, as opposed to a massless particle) and is related to its momentum through the Planck constant. According to De Broglie, a wave is associated with each moving particle, which is called a 'matter wave'. This wave has a wavelength. Matter waves bear certain characteristics: (1) The lighter the particle is, the greater the De Broglie wavelength will be. (2) The faster the particle moves, the smaller is its De Broglie wavelength. (3) The De Broglie wavelength of a particle is independent of the charge or nature of the particle. And (4) matter waves are not electromagnetic. Electromagnetic waves are produced only by a charged particle. Matter waves are of two types—longitudinal and transversal. Transverse waves are like those on water, with the surface going up and down. On the other hand, longitudinal waves are like those of sound, consisting of alternating compressions and rarefactions in a medium.

By analogy with the wave and particle behavior of light that had already been established experimentally, particles might have wave properties in addition to particle properties. The objects of everyday experiments, however, have a computed wavelength much smaller than that of electrons; their wave properties have never been detected as familiar objects, showing only particle behavior. De Broglie waves play an appreciable role, but only in the realm of subatomic particles. De Broglie waves account for the appearance of subatomic particles at conventionally unexpected sites because their waves penetrate barriers much as sound passes through walls. Thus a heavy atomic nucleus occasionally can eject a piece of itself in a process called 'alpha decay'. The piece of the nucleus (the alpha particle) has insufficient energy as a particle to overcome the force barrier surrounding the nucleus, but as a wave, it can leak through the barrier; that is, it has a finite probability of being found outside the nucleus. These phenomena can be explained by quantum tunneling.

Quantum Tunneling

Quantum tunneling is the quantum mechanical phenomenon where wavefunction can propagate through a potential barrier. The transmission through the barrier can be finite and depends exponentially on the height and width of the barrier. The wavefunction may disappear on one side and reappear on the other side. The wavefunction and its first derivative are continuous. In a steady state, the probability flux in the following direction is spatially uniform. No particle or wave is lost. Tunneling occurs with barriers of thickness around 1–3 nanometers and smaller. Tunneling is a quantum mechanical effect. A tunneling current occurs when electrons move through a barrier that they classically should not be able to move through. Quantum mechanics tells us that electrons have both wave- and particle-like properties. Tunneling is an effect of wave-like nature. Quantum tunneling happens only over extremely tiny, microscopic distances and is unrelated to the entanglement of teleportation. It has to do with the fact that elementary particles are waves, so they are more spread out than classical particles.

Quantum tunneling falls under the domain of quantum mechanics: the study of what happens at the quantum scale. Tunneling cannot be directly perceived. Much of its understanding can be shaped by the microscopic world, which cannot be explained by classical mechanics. To understand the phenomenon, particles attempting to travel across a potential barrier can be compared to a ball trying to roll over a hill. Quantum tunneling

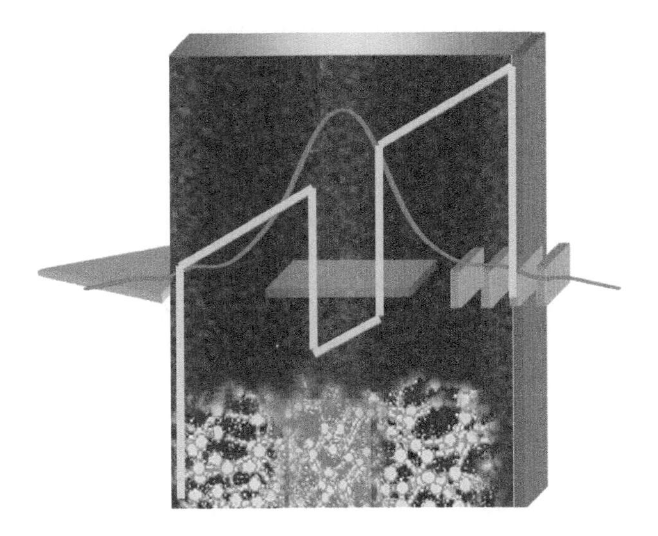

FIGURE 8.1
Quantum tunneling through a double barrier system made of disordered carbon. The wavefunction decays from the center of the well. The barriers are bent when a potential is applied to them.

can be explained in terms of the Heisenberg uncertainty principle in that, in general, a quantum object can be known as a wave or as a particle. In other words, uncertainty in the exact location of light particles allows these particles to break the rules of classical mechanics and move in space without passing over the potential energy barrier. Quantum tunneling may be one of the mechanisms of proton decay.

The Tunneling Problem

The wavefunction of a particle summarizes everything that can be known about a physical system. Therefore, problems in quantum mechanics analyze the system's wave function. Using the equation, the wavefunction can be deduced. The square of the absolute value of the wavefunction is directly related to the probability distribution of the particle's position, which describes the probability that the particle is at any given place. The wider the barrier and the higher the barrier energy are, the lower the probability of tunneling. A simple model of a tunneling barrier such as the rectangular barrier can be analyzed and solved algebraically. In canonical field theory, the tunneling is described by a wavefunction that has nonzero amplitude inside the tunnel but zero current because the relative phase of the amplitude of the conjugate wavefunction (the time-derivative) is orthogonal to it.

Tunneling means seeing the same wavefunction on the opposite side of an active barrier that lies as a constant between two active elements, like Satwa between Rajas and Tamas, that interact with each other and find a similarity. Attraction becomes repulsion, negative becomes positive, etc. We shall discuss the details of the tunneling process shortly. Maya-Shakti wants two opposite states or different characteristics together to show that she can control them like two electrons of opposite spins (or attractive/repulsive forces or electric and magnetic fields). They can stay in one state well. But what happens if something comes out of the reaction? Maya-Shakti does not like three similar things existing in one level or state since it will break the reaction and produce chaos. So the third element has to be different from the first two, and we can have as many particles of the third kind as possible. These are called Boson particles if they are produced from fermions that have zero mass. So they can be in the upper layer like a cover for the two fermions, and they can spread on the surface of the sphere. They will form the space or vacuum or Akash. From the electric field (E) and magnetic field (M) light, a Satwa element, light, is formed. From Rajas (E) and Tamas (m), a constant c^2 in $E= mc^2$ is explained. Swata is defined as a force ($\boldsymbol{F}$), which is equivalent to knowledge. From a dual state P and V, temperature (a boson) is created through the equation $pv = nRT$. In the formation of Boyle, the applied pressure reduces the volume. The low pressure and large volume (space) can live comfortably together without changing the temperature. If we squeeze the space, it increases pressure, and to keep both P and V happy, the temperature must change. This leads to an increase in the number of bosons and to a blast that may create a frequency to respond to the change. So the heat will be transferred from the center of a cavity to the surface, creating a sky. A cosmic expression from the duality can create as many stars as possible. So the zero of the third kind is different from the two others, which are interesting and produce the third one.

So how does tunneling takes place when the resistance works against the curved or spherical wavefunction? We need to know the orientation of two interacting elements. They are opposite, meaning that they are normal to each other, like Kali on Shiva, Laxmi on Vishnu (lying flat), or Brahma on Vishnu. It follows Pythagoras's theorem $c^2= a^2 + b^2$, where a and b are perpendicular to each other as the height of a pyramid and its shadow on the ground. This equation can be derived from Sankhya philosophy. Since the height remains constant, the shadow adjusts its length with the alignment of the Sun. So the Sun is interacting as Rajas with the pyramid (Satwa) with the shadow (Tamas). Now we take two circles representing a and b, which are normal to each other like electric and magnetic field vectors. Light is ejected normal to the axes of the circles, and the speed will be constantly adjusted by the interaction of the circles. In atomic physics, electrons and holes in the hand recombine and produce light normal to the hands, as leaves of a plant grow normal to the branches and can spread out in space. The gravitational field is produced radially outward of the Sun, which may be a result of Yukawa and Coulomb's potential. In the tunneling process, mind is tunneling through the potential barrier (knowledge) and produces life. When a particle-like mind reaches the barrier, it creates an antiparticle inside the barrier, and they start interacting since on the other side of the barrier the same particle is formed as an image of the original particle. This particle and antiparticle interfere and produce an interference spectrum, as observed in the tunnel spectra.

This is like Vishnu, who acts as an inactive barrier to the water, and the water vapor is supposed to overcome the resistance of Vishnu by the heat produced by Maya-Shakti. The body of Vishnu is divided to pass through the tunnel current. In practice by applying a potential across the potential barrier (V), it is broken into $V_1V_2V_3$ ... so that $V = V_1 + V_2 + V_3 + V_4$ ... . As the current can pass through different potentials V_1, V_2 etc., they unite on the other side of the barrier and recombine. So the incident wavefunction of the total potential barrier is divided in the process. Now the wave can change its phase in the barrier. This is like a flip of spin or self-spin. This can happen since the goddess Maya-Shakti is the source of heat energy, i.e., Prana (life), and controls the entire tunneling process. Now the wave can change phase in the barrier. This is like a flip of spin or self-spin. This can happen since the goddess Maya-Shakti is the source of heat energy that is Prana (life) and controls the entire tunneling process. For example, the two aspects of the particle–antiparticle cancel each other at the interference point (or region), and it produces the heat energy inside the barrier and then quickly transfers it to the other side of the barrier depending on the size of the barrier. Mind is the neutralization of the particle–antiparticle, knowledge is the transformation of particle–antiparticle, and life is energy transfer. The Prana of heat energy moves along the surface of the barrier and then is emitted like heat on the surface of Vishnu. The mechanism or principle of tunneling includes (1) no return of the particle; (2) the particle can die if tunneling does not work; (3) the particle does not hesitate to tunnel as it is free from fear; (4) it can face uncertainty; (5) it makes the maximum effort before jumping; (6) the probability of success must be maximum, i.e., it will take a minimum time to tunnel; (7) tunneling is the equivalent of death of the original particle, (8) only the maximum amount of energy is transferred, and (9) the energy gap is filled with sound or waves that carry the energy. So the goddess Maya-Shakti divides the tunnel barrier into two halves that are parallel to each other and normal to the direction of tunnelling. She places the antiparticle and then the particle on each side. The particle–antiparticle interaction produces an interference pattern and also the tunneling spectrum.

Next, we shall explain the features of superconductivity since this chapter deals with the tunneling of particles in pairs, called the Cooper pair.

Superconductivity

Superconductivity is a set of physical properties observed in certain materials where electrical resistance vanishes, and magnetic flux fields are expelled from the material when cooled below a certain (transition) temperature. Superconductors and superconducting

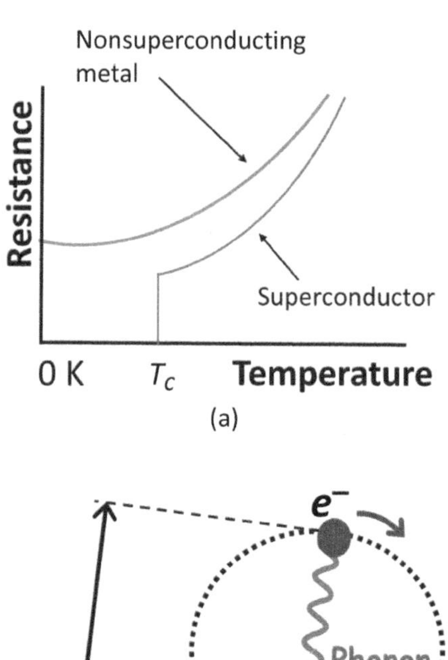

FIGURE 8.2
(a) Superconducting transition of a superconductor showing zero resistance below a transition temperature T_C instead of a normal metal. (b) Two electrons form a Cooper pair however, they are separated by a wavy line representing a vibrational mode, called phonon.

materials are metals, ceramics, organic materials, or heavily doped semiconductors that conduct electricity without any resistance so that electrons can be moved without losing any energy to heat. An electron has a charge and a spin. The spin of an electron can connect to another electron's spin, and the same applies to the atomic lattice in which the electrons are located. The atoms can also move and can thus actually give rise to superconductivity through their lattice fluctuations.

Superconductors explain the magnetic field from the interior by setting up an electric current at the surface. The surface current creates a magnetic field that exactly cancels the external magnetic field. This electric current at the surface of the superconductor appears at a temperature that is less than the critical temperature, so that B + A inside the superconductor. Below a certain critical temperature, materials undergo a transition into a superconducting state, characterized by two basic properties: First, no resistance is offered to the passage of electric current. When resistance falls to zero, a current can circulate inside the material without any dissipation of energy. Second, if sufficiently weak, external magnetic fields will not penetrate the superconductor but remain at its surface. The use of superconductors in magnets is limited by the fact that strong magnetic fields (or a large current) above a certain critical value, depending on the material, cause a superconductor to revert to its normal or nonsuperconducting state, even though the material is kept well below the transitional temperature.

In 1950, it was clearly shown for the first time that the theory of superconductivity must take into account the fact that free electrons in a crystal are influenced by the vibrations of atoms that define the crystal structure, called the 'lattice vibrations'. In 1953, in an analysis of the thermal conductivity of superconductors, it was recognized that the distribution of energies of the free electrons in a superconductor is not uniform but has a separation called the 'energy gap'.

The BCS theory (Bardeen–Cooper–Schrieffer theory, named after scientists John Bardeen, Leon Cooper, and John Robert Schrieffer), proposed in 1957, was the first microscopic theory of superconductivity since Heike Kamerlingh Onnes's discovery in 1911. BCS theory describes superconductivity as a microscopic effect caused by a condensation of Cooper pairs. The theory is also used in nuclear physics to describe the pairing interaction between nucleons in an atomic nucleus.

At sufficiently low temperatures, electrons near the Fermi surface become unstable against the formation of Cooper pairs. Cooper showed that such binding will occur in the presence of an attractive potential, no matter how weak. In conventional superconductors, an attraction is generally attributed to an electron–lattice interaction. The BCS theory, however, requires only that the potential be attractive, regardless of its origin. In the BCS framework, superconductivity is a macroscopic effect that results from the condensation of Cooper pairs. These have some Bosonic properties and bosons at sufficiently low temperature, forming a large Bose–Einstein condensate. In many superconductors, the attractive interaction between electrons (necessary for pairing) is brought about indirectly by the interaction between the electrons and the vibrating crystal lattice (the phonons).

The original results of the BCS theory described an *s*-wave (having the net spin of zero) superconducting state, which is the rule among low-temperature superconductors but is not realized in many unconventional superconductors such as the *d*-wave (a spin-triplet state with net spin = 1) high-temperature superconductors. Extensions of the BCS theory exist to describe these other cases, although they are insufficient to completely describe the observed features of high-temperature superconductivity.

BCS theory derived several important theoretical predictions that are independent of the details of the interaction since the quantitative predictions hold for any sufficiently weak attraction between the electrons; the condition is fulfilled for many low-temperature superconductors—the so-called weak coupling case. The electrons are bound into Cooper pairs, and these pairs are correlated due to the Pauli exclusion principle for the electrons, for which they are constructed. Therefore, to break a pair, one has to change the energies of all other pairs. This means that there is an energy gap for single-particle excitation, unlike in normal metal (where the state of an electron can be changed by adding an arbitrarily small amount of energy). This energy gap is highest at low temperatures but vanishes at the transition temperature when superconductivity ceases to exist. The BCS theory gives an expression that shows how the gap grows with the strength of the attractive interaction and the normal phase single-particle density of states at the Fermi level. Furthermore, it describes how the density of states is changed on entering the superconducting state, where there are no electronic states anymore at the Fermi level. The energy gap is more directly observed in tunneling experiments and the reflection of microwaves from superconductors. BCS theory predicts the dependence of the value of the energy gap at a temperature over the critical temperature. The ratio between the value of the energy gap at zero temperature and the value of superconducting transition temperature, expressed in energy units, takes the universal value ($T = 0$) = 1.764 units, independent of material. Due to the energy gap, the specific heat of the superconductor is suppressed strongly (exponentially) at low temperatures, there being no thermal excitations left. The energy gap behaves like the Purusha, which plays the most crucial role in tunneling between two superconductors.

Josephson Junction

In the year 1962, scientist B. Josephson predicted that two superconductors, separated by a barrier consisting of a thin insulator, should be able to allow an electric current to pass between them due to the tunneling of Cooper pairs. This phenomenon is known as the 'Josephson effect' or 'Josephson tunneling'. Josephson junction (a quantum mechanical device) is made up of two superconducting electrodes, separated by a barrier (a thin insulating tunnel barrier, normal metal, semiconductor, ferromagnet), which results in sandwiching a thin nonsuperconducting layer, such that the electrons can tunnel through the barrier. The coherence of the wavefunction in the superconductor leads to DC or AC currents. In AC currents, a constant chemical potential difference (voltage) is applied on the Josephson effect, causing an oscillating current to flow through the barrier. On the other hand, in the DC Josephson effect, a small constant current is applied, resulting in a constant supercurrent flowing through the barrier. A Josephson junction consists of a localized discontinuity or weak link in the order parameter of two superconducting electrodes, where the dissipationless current ruled by the Cooper pairs transport is controlled by the macroscopic quantum phase difference across the junction. Josephson junction is a nonlinear element because it combines negligible dissipation with extremely large nonlinearity: The change of the qubit state by only one photon in energy can modify the junction inductance by the order of unity. The presence of magnetic fields near the superconductors influences the Josephson effect, allowing it to be used to measure very weak magnetic fields.

Quantum tunneling between two metals needs a thin layer of the in-between insulator. A bandgap should be created between the metals having no bandgap. A superconductor has a very small energy gap. Cooper pair tunneling between two superconductors separated by a metal or an insulator gives a very interesting tunneling phenomenon, called Josephson tunneling. The property of the Josephson junction can be tuned by the presence of a magnetic material or spin center that produces a spin-triplet state. A bound state can

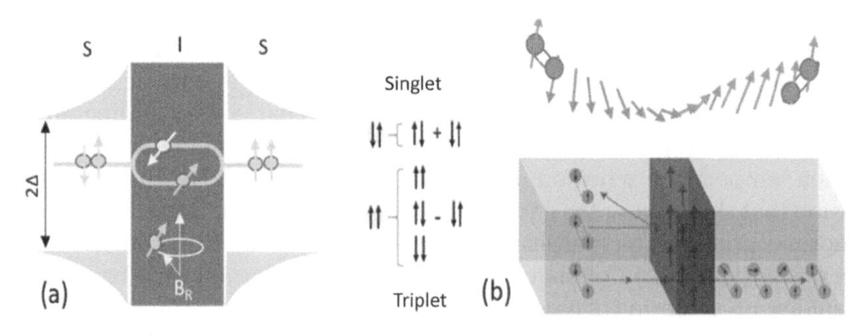

FIGURE 8.3
The resonant transport features can be evaluated as a spin-polarized Andreev bound state scenario due to spin precession caused by the Rashba field at the interface. It also explains the formation of a long-range triplet capable of moving beyond the single grain regime. Considering our verification of the RSOC at the grain interface, spin-polarized Andreev Bound states can give rise to this phenomenon. This can be expressed as a pair of electrons having opposite spins are tunneling through a barrier. The net spin is zero for the pair. One of the spins is flipped during this process and two spins become parallel that produces a spin 1 or a triplet state.

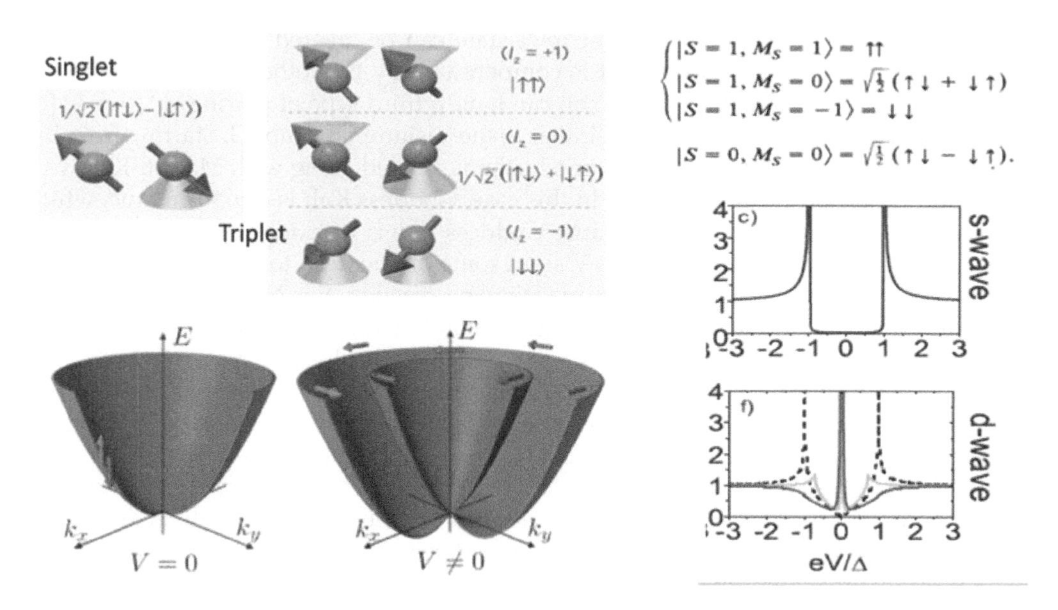

FIGURE 8.4
Electrons have spin states of +1/2 and -1/2. Therefore, there are four possible combinations: (1) Up, Up; (2) Down, Down; (3) Up, Down; and (4) Down, Up. For triplet states, the angular momentum sums to 1 for singlet states, the angular momentum sums to 0 triplet (1, = +1). In a spin singlet state two spins are pointing in opposite directions yielding a net spin of zero. In a spin triplet state, both spins can be up or down direction yielding a net spin state 1 or -1 in addition to a spin zero state. For spin-orbit coupling the potential well is split into two parts like a cup inside another cup. This yields a conductance peak at zero energy for a spin triplet state instead of an energy gap for a spin singlet state.

This phenomenon can be explained by the battle of two cycles.

be formed in the junction that gives Andreev reflections. This junction is the key to fabricating artificial atoms, or qubits, based on the supercurrent through the junction.

The tunneling through the superconducting junction supports the duality (or uncertainty) concept. It is the playground of all exotic quantum phenomena. Like the Bose–Einstein condensate, I find that the superconductivity can be linked with the Sankhya philosophy for the ground state, which behaves like the Purusha as a dormant state (*not disturbed by the weak magnetic field or the applied current*). It rejects the applied magnetic field. Second, the electrons form a pair that arises from dualism. Third, in the presence of a magnetic or a spin center, the singlet state is transformed into a spin-triplet state. A gapped superconductor can be transformed into a gapless superconductor, which shows some very interesting tunneling properties. Even though there are extensive studies of spin-triplet superconductors, an exact mechanism for singlet-to-triplet transition has not been established. Hence the pairing mechanism of high-temperature superconductors remains unsolved. In this chapter, I shall explain the features of spin-triplet superconductors.

Based on this simple logic or law of the Universe, Pauli's exclusion principle can be reexplained. In one state two electrons of opposite spins can be present, like (↑↓). The half spins have equal mass but opposite polarity. They can have an exchange interaction and as many as possible bosons (zero or integer spins) in one level, just as many stars are allowed in the sky but one Sun (Brahman) and Maya (its heat or other characteristics). Planets in different orbits are allowed.

From a spin-singlet (↑↓) state a spin-triplet state can be created where two electrons of equal polarity can exist; however, it is compensated by two other spins of equal and opposite polarity, like |↑↑> and |↓↓>. You can have a third type of pairing, |↑↓> + |↓↑>, which is similar to the subject. It is also similar to the Vishnumaya model. Starting from the bosonic bath of Vishnu, Brahma (light) or a boson is created along with Madhu-Kaitava's body, which is massive and fermionic. In this case, Goddess Kali is also fermionic, which is the resistive force to the process. Because Goddess Kali is working on Vishnu, this conversion takes place. Goddess Kali gives Vishnu some momentum to rise (and then wakes it). This resistive power is the most important in any reaction. For Newtonian mechanics, it is inertia (m); in Boyle's law, it is volume change; in Einstein's equation, it is m, which is acting as a barrier against tunneling. It is a gap of energy or physical space. To produce a triplet state, a barrier of ferromagnetic material is needed since the triplet has an antiferromagnetic behavior (↑↓). Maya-Shakti in a special form converts the union of Shiv-Shakti (singlet) into three triplets |↑↑>, |↓↓>, and |↑↓> + |↓↑> and moves these states in the sky from the water levels as clouds.

This picture becomes interesting with the addition of the second potential barrier. We believe any Universe in an elapsed shell can be modeled as a double barrier system where Maya-Shakti is protecting the world full of water and life with the two barriers. Heat or light tunnels through one of the barriers and resonates waves on Earth and then leaves through the other barrier. The self-spin picture works very well as a resonance in the well. This self-spinning of particle or resonance can be described as a rebirth. The tunnel means great attraction, which gives liberation of the particle. It happens from the perfect matching of holes and electrons like filled and unfilled or sufficient and deficient. Find similarities between the two, and they have similar energy. Reentrant superconductivity can be explained by tunneling between two vortices. Assuming the vortices behave like black holes, tunneling between them can be explained by a wormhole. The transition from the black holes to wormhole at a temperature can be compared to the transition from a

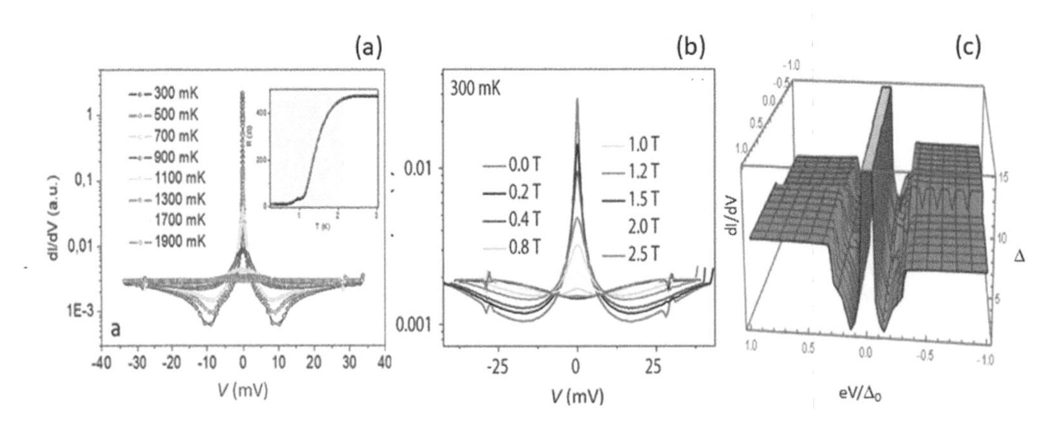

FIGURE 8.5
(a) The zero-energy peak is highly dependent on temperature and is only observed at temperatures below the superconductivity transition onset, as indicated in the inset. The peak height decreases exponentially with increasing temperature, and FWHM increases exponentially. **(b)** The ZBCP does not split up until it is completely suppressed and inverted to a V-shape dip, like what has been observed in other triplet superconductors. The field strength where this inversion of the peak occurs marks the crossover from a superconducting to an insulating regime, as shown in the high field magnetoresistance data. **(c)** 3D plots of dI/dV with the variation eV/zero energy gap (Δ_0) for different values of the superconducting gap (Δ) can explain the experimental data.

gapless superconductor to a gapped superconductor (vortices) that are connected by a tunnel junction. We shall understand the wormhole state in detail in Chapter 9, which can explain the features of spin-triplet superconductors.

Conventional (s-wave) superconductors are routinely used in fabricating coupled qubits for quantum processing technologies. However, unconventional (spin-triplet) superconductors are expected to offer advantages as basic elements of topological quantum computers for complex operations since the symmetry of the order parameter (Δ) corresponds to a doubly degenerate chiral state.

Cooper pairs conducting through this atomically thin layer would be subjected to a nonequivalent electric field (defined by the interface potential as a step function), and thus a spin-orbit coupling of the Rashba-type (RSOC) followed by spin-triplet superconductivity arises. In Edelstein's model, the interface was described as a double electric layer, which is similar to a vortex structure. However, details of the microstructure of the interface and their interconnectivity remain unclear until today.

By applying an electric field in the presence of RSOC, a nonequilibrium magnetization or spin accumulation in the interface takes place with the generation of spin-polarized currents, which can form an effective S-F-S structure. A quasiclassical theory describing the interface of superconductor and ferromagnets in the presence of the SOC has also been developed. All these problems can be solved with the proper understanding of the originally stated Fulde–Ferrel and Larkin–Ovchinnikov (FFLO)-like state in the 1960s with the modulated order parameter; however, it was based on the *d*-wave superconductor. The symmetry of pair wavefunction is given by momentum ⊗ spin ⊗ frequency. A key parameter to describe the symmetry breaking of the grains producing the grain boundary region is defined as $\Sigma = [C1\cdot(C2\times C3)]/[a\cdot(b\times c)]$, which finds similarity with symmetry breaking of the angular momentum vector in introducing the helicity, popularly known as the Rashba SOI described by Edelstein as interface SO: $H_S = \alpha\ (\rho \times \boldsymbol{c}) \cdot \boldsymbol{\sigma}\ (\boldsymbol{c} \cdot \boldsymbol{r})$, where $\boldsymbol{c}$ is one of the two nonequivalent normal unit vectors, and the $\boldsymbol{\sigma}$ function describes the interface potential with a position vector ($\boldsymbol{r}$).

Spin-orbit coupling enables the flow of angular momentum between the spin angular momentum of the electronic system and the mechanical angular momentum of the lattice. This provides the opportunity for more energy-efficient electrical manipulation of magnetic order. The spin-orbit coupling is an important measurable in the carbon nanostructures, including the nanotubular part of the graphitic wormhole. Spin-orbit coupling is presented for the fermions located in exotic graphene structures as graphene is a wormhole as per the biological systems. Considering the influence, the two-component Dirac equation is changed into the usual four-component form. As a consequence, chiral fermions should be detected close to the wormhole bridge. Again, electric dipole spin resonance (EDSR) is the coupling of the electron spin with an oscillating electric field. Similarly to electron spin resonance, in which electrons can be excited with an electromagnetic wave with the energy given by the Zeeman effect, in EDSR the resonance can be achieved if the frequency is related to the band splitting given by spin-orbit coupling in solids. While in ESR the coupling is obtained via the magnetic part of the EM wave with the electron magnetic moment, the EDSR is the coupling of the electrical part with the spin and motion of the electrons. This mechanism has been proposed to control electrons' spin in quantum dots and other mesoscopic systems.

Qubit

Superconducting devices can be used to explore the boundaries between the quantum and classical worlds and could also have applications in quantum information. The quantum

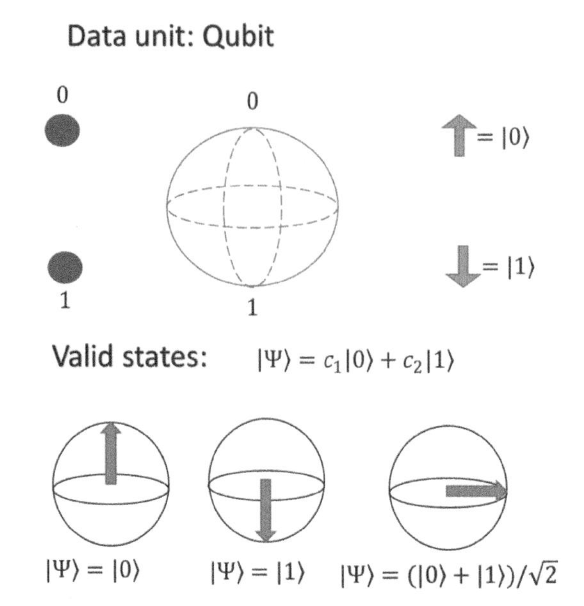

FIGURE 8.6
Unlike 0 and 1 bits, a qubit can be described as a superposition of two bits or two energy states. The arrows show the states schematically. A qubit can be represented as an atom. A qubit can be imagined as a sphere with an arrow pointing from the center to the surface of the sphere. The state of a qubit can be realized by the direction of the arrow e.g., up and down arrows represent zero and 1 state, respectively.

world looks very different from the ordinary world: A quantum particle can exist simultaneously in two places, while its speed and position can be measured with complete accuracy at the same time. Moreover, its mass is small enough that a quantum particle can tunnel through energy barriers, such that its classical counterparts could never cross. Recently, some micron-sized objects have been discovered using standard semiconductor fabrication techniques. Objects that are small on everyday scales but large compared with atoms can also behave like quantum particles. These artificial quantum objects are used as quantum bits in a quantum computer, which could perform certain computational tasks much faster than any classical computing device. They are called 'qubits' since they are based on a two-level quantum system that can be useful to test the laws of quantum mechanics.

Quantum mechanics is based on an uncertainty principle, where the fundamental particles are nearly invisible. An ideal place to apply these mechanics is in an atom that consists of negatively charged particles revolving around a center consisting of positively charged particles, leaving the space between these two opposite particles empty. An artificial atom can be made by superconducting (zero electrical resistance) materials that are popularly known as a qubit in the form of a closed loop of superconductor with a small gap of nonsuperconducting material. The supercurrent flows in the loop in both the clockwise and the counterclockwise directions, and a superposition of states can be created that is equivalent to the creation of a void. In a quantum system, duality works in the qubit, which is an oscillator. Microwave is the input that gets converted into different frequencies/channels after interaction with qubits. Photons pass through different stages of wheels, which increases the fidelity of the circuit. Photons can have different colors or strings that can be combined to make a cloth of different designs and textures. Photons are the elements that, like strings, can be braided to store more information at the joints of strings. It is like a hologram starting from a spinning machine producing circles that mix

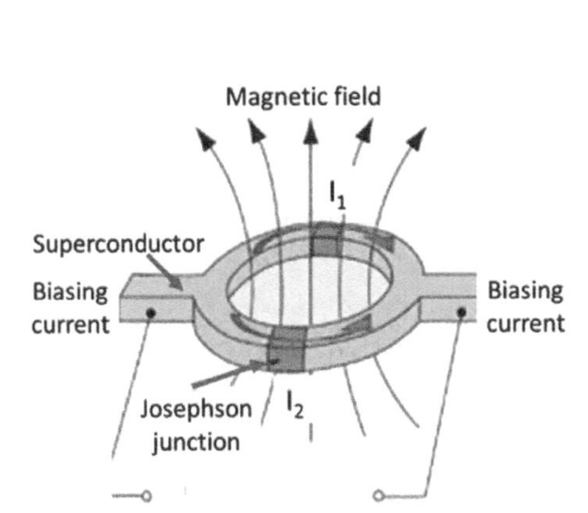

FIGURE 8.7
(a) Superconducting quantum interference device (SQUID) structures produces quantized magnetic fields. A SQUID loop looks like a ring with Josephson junctions. The supercurrent can flow in two opposite directions in the loop which also produces quantized magnetic flux. (b) Battle of two cycles. Interactions between two gods and a goddess. One god is spinning the body of his wife, a goddess. Another god is using a wheel like device to cut the body of the goddess into pieces. The rotations create objects and fine on earth and the planets in the sky.

and produce multiple colors. This process needs the intelligence of a multichannel system that produces different colors. With the arrangement of several connected artificial atoms, a quantum computer has been developed that can outperform a conventional supercomputer in terms of complex operations. Although, in theory, one can develop artificial intelligence based on these qubits, this has not been demonstrated yet. Many people believe that quantum mechanics works like an illusion created from duality maintaining a competition of two opposite things and on the principle of superposition of states. Maya can be seen as a black cover against the white uncovered object where our attention can be focused. It is analogous to a male face covered with a beard or a female face with a headcover. Just the center part looks incompletely covered with Maya, and we can imagine Brahma and Maya together. Imagine you see either Maya or Brahma or the partial nature of either of them. This is like a classical bit, 0 or 1. In quantum, they are indifferent. We are finding the elementary vibration that gives all colors, and finally we see all colors with the elementary vibrations.

Pauli Spin Matrices

A set of three Pauli's spinors (σ_x, σ_y, σ_z) represent inversion, rotational, and translational symmetry like up–down, clockwise–counterclockwise, and the mirror image for spins. Multiplications of two of the spinors yield the third spinor, which reminds me of the three Gunas described in Sankhya philosophy. Mirror image is like Maya, which can be described by Tamas, whereas the rotation is described as Rajas. Three spinors in combination give an 'identity' spinor or no change, as observed from the successive operations of each spinor. These operations can be explained by the maximally crossed diagrams as the propagation of Cooperons and the backscattering process. Pauli spin matrices represent a spin angular momentum operator for half-integer spin particles. In three–dimensional space, they are σ_x, σ_y, and σ_z (or σ_1, σ_2, and σ_3) and I, and they follow several relationships or commutation rules as follows:

$$\vec{\sigma} = \sigma_1 \hat{x}_1 + \sigma_2 \hat{x}_2 + \sigma_3 \hat{x}_3 \sigma_1^2 = \sigma_2^2 = \sigma_3^2 = -i\sigma_1\sigma_2\sigma_3 = \begin{pmatrix} 1 & 0 \\ 0 & 1 \end{pmatrix} = I\left\{\sigma_j, \sigma_k\right\} = 2\delta_{jk} I$$

So far we have explained two types of particles, bosons and fermions, having integer and half-integer spins and there is always competition:

$$\left[\sigma_j, \sigma_k\right] = 2i\varepsilon_{jkl}\sigma_l \left(\vec{a}\cdot\vec{\sigma}\right)\left(\vec{b}\cdot\vec{\sigma}\right) = \left(\vec{a}\cdot\vec{b}\right)I + i\left(\vec{a}\times\vec{b}\right)\cdot\vec{\sigma} \;\; \sigma_j\sigma_k = \delta_{jk}I + i\varepsilon_{jkl}\sigma_l$$

between these two particles. How do we combine them in a system, or how do we make conversion between these two particles? A spin chain like a magnet does not have an energy gap that can resist the superconducting system with a small energy gap. The spin-half particles can be split into two halves or spinors. A pair of spinors that are duals can open up a bandgap. In other words, the spin half particles can be described by $\sigma_x + i\sigma_y$ for one- or two-dimensional chains. For the third dimension, another spinor σ_z can be added to the system. These three spinors σ_x, σ_y, and σ_z, can be associated with another spinor, call the identity spinor 1. The combination of spinors that are opposite to each other (duals) can be arranged in a cyclic order, e.g., $[\sigma_x, \sigma_y] = 2i\sigma_z$, $[\sigma_x, \sigma_z] = -2i\sigma_y$, and $[\sigma_y, \sigma_y] = 2i\sigma_x$. A combination of two of the three spinors produces the third spinor as σ_x, σ_y, and σ_z. These relationships remind me of the three Gunas described in Sankhya philosophy that also follow a cyclic order. Pure white light is produced from the combination of three colors (three Gunas) as the identity spinor is produced from σ_x, σ_y, and σ_z. The identity spinor can be described as Purusha and the three others as Prakriti. In practice, the spinors can describe the state of polarization of light or photon, i.e., vertical and horizontal (or the left and the right) states. A beam of light can be split into two halves by using a polarizer that can rejoin. This two-level operation forms a qubit and can be described as the rotation of a circle. In general, a sphere (called a Bloch sphere), having three axes of rotations, can describe the full operation of a qubit by using three spinors σx, σy, and σz. A qubit operates based on the superposition of two states, 0 and 1. The 0 and 1 states can be compared to inactive or static Purusha and to the dynamic Prakriti, respectively. They can also be compared to orbit (or charge) and spin, respectively. The superposition of orbit and spin form a triplet state where two spins are aligned in the same direction or combined in such a manner to produce an integer, or $S = 1$ spin. The spin can be expressed by two spinors as $\sigma_x + i\sigma_y$. The superposition is described as the duality of Purusha and Prakriti as the foundation of Sankhya philosophy, which yields three spinors as three Gunas. The fermionic operator can be converted into a tensor product of the spinors.

Similarly, bosonic operators can be mapped by spinor $\sigma\pm$ states so that the whole condensate with the magnetic impurities can be described by the three spinors. The expression for the total energy or the Hamiltonian can be expressed by the three spinors. As a result, the spinors can describe the quantum mechanical problem. Here, we can explain the tunneling problem in terms of qubits.

An application of these quantum phenomena will be quantum vortices that are observed in superconducting materials, followed by the tunneling between the vortices and forming a wormhole structure at the end of the chapter (Refer to Figure 8.6).

Summary

We have introduced a new approach to quantum mechanics based on the quantum vortex model that arises from the dualism model. This chapter deals with the most important applications of quantum mechanics, i.e., tunneling, in addition to wave-particle duality. We showed the application of quantum mechanics in spin-triplet superconductivity, which is based on the tunneling of the superconducting phase. We then reviewed superconductivity and the Josephson junction and showed the application in qubits. We gave an in-depth analysis of quantum tunneling based on dualistic philosophy that can be applied to wormholes. This will be discussed in Chapter 9 in the context of space–time distortion.

9

Quantum Technology and Space–Time Distortion: Black Holes, Condensates, and Quantum Devices

- **Space–time distortions and disorder (noise)**
- **Black holes and holographic Universe**
- **The SYK model of a maximally chaotic system**
- **Quantum simulations of black holes and wormholes**

In this chapter, we shall initiate a discussion that relates to Chapter 1. In the beginning, everything was motionless, but space is split into two kinds of (periodic or aperiodic) cycles, which complement each other forever. When the speed of one cycle is intensified, the traversed area of the other one is affected. These two cycles work perpendicularly to each other. They can be in phase or antiphase. However, one could stop one of the cycles by closing the area that tends to zero. Once one of the cycles reaches singularity, the other can also be stopped. However, before stopping, a burst of energy is emitted normal to the area of the first cycle signifying creation. This process is like an expression of a smile appearing on the face after the creator becomes extremely angry. The space can be split again and again after releasing the energy as the harmonic of his laughter resonates and creates the sound of all frequencies. When two of the cycles are in phase, a bound state is formed. Increasing the speed of one cycle reduces the speed of the other to a point where a flip takes place. Having crossed that point, one cycle enhances the intensity of the other. In this way, the bound state can be overcome, and waves/particles can propagate freely. The transition point of the bound to the free state is creation. The two cycles are space-like and time-like, producing the charge (or the mass) and the phase (or a magnetic field) or spin, respectively. Through the distortion of the space–time 'creation' is initiated locally. At the transition point, space–time distortion vanishes as the singularity is reached. It gives a strong resonance peak (like my smile). In the quantum world, this is called the zero-bias conductance peak, which can also be associated with two satellite peaks. These three peaks correspond to a spin-triplet state consisting of |↑↑>, |↓↓> and their superposition |↑↓> + |↓↑>. From any dual motion of cycles, particle-wave and their duality as matter wave (explained in Chapter 8) is created.

Regarding duality, Sankhya philosophy explains not only the space–time (Purusha–Prakriti) but its distortion (Vikriti), which explains a process that adds noise to the system. For example, the interior of a black hole (or a quantum vortex) is maximally chaotic and should be controlled to achieve the stability of the entire system. A holistic model for this control is described in Sankhya philosophy (in Chapter 4). Here we show, in condensed

DOI: 10.1201/9781003304814-10

matter physics, how to model disorder in a vortex that represents a black hole. Let us discuss black holes briefly.

Recently published was an image of a black hole in our galaxy that was produced not only by an extremely high-quality telescope but also by very advanced image processing. The interior of a black hole can be understood by theoretical modeling only. A philosophical understanding would be useful for the quantum simulation of a black hole. For the simulation of the black hole structure, the inputs will be Satwa–Rajas–Tamas or static-dynamics-massive characteristics. In Chapter 1, we described a vortex, an infinitely long tunnel, and a horizon. We discussed how space could be created. The space was created from a point or an infinitely deep potential. The end of the tunnel is not empty but filled with the densest materials that have a maximally chaotic structure. According to Buddhist philosophy, creation starts from 'nothing' or an empty state. However, Hindu philosophy suggests that creation started from a state that is absolutely filled. This is a 'joyful state' that does not show any sign of disturbance or excitement. This state compares with the core of a black hole, which is completely filled as it is a most compact object. The massive core of a black hole represents a static structure. Philosophically, this has a characteristic of Purusha, which is a dormant state. A black hole can have the energy of energy, which also represents the Prakriti. The maximally chaotic state represents Tamas. The attractive nature of black holes is attributed to Rajas. The massive black hole creates the space–time distortion.

In the Chapters 1 and 2, we introduced the concepts of the creation of *matter* and *mind* and of primordial *energy* from a philosophical as well as a spiritual point of view. In this chapter, we give a parallel interpretation of creation through space–time and one of the fundamental forces, i.e., gravitation. We have addressed this point through Space–time distortion (S-T D) in the vortices observed in superconductors (Chapter 8). The unification of space and time originally proposed by Einstein applies to black holes (and connected black holes or wormholes) where the S-T D vanishes. A vortex model is developed based on the black holes where space (and time) become distorted as given by the Anti–de Sitter (AdS) space. However, S-T D phenomena can be simulated by microresonator rings, by quantum dots, and by a set of superconducting qubits, as discussed in this chapter.

Let us understand the terms 'space–time equivalence' and 'space–time distortion', followed by black holes and their connectivity in terms of a model of the disorder.

Space–Time

In physics, space–time is a mathematical model that fuses the three dimensions of space and the one dimension of time into a single four-dimensional manifold. Space–time diagrams can be used to visualize relativistic effects such as why different observers perceive differently where and when events occur. Until the 20th century, it was assumed that the three-dimensional oval geometry of the Universe (its spatial expression in terms of coordinates, distances, and directions) was independent of one-dimensional time. It was Albert Einstein who helped to develop the idea of space–time as part of his theory of relativity.

In four-dimensional space–time, the analog to distance is an interval. Although time comes in as a fourth dimension, it is treated differently than spatial dimensions. The fundamental reason for merging space and time into space–time is that space and time are separately not invariant, which is to say that, under the proper conditions, different observers will disagree on the length of time between two events (because of length contraction). Special relativity provides a new invariant, called the 'space–time interval', which combines distances in space and time. All observers who measure time and distance between any two events will end up computing the same space–time interval. A space–time diagram is typically drawn with only a single space and a single time coordinate. To gain insight into how space–time coordinates are measured by observers in different reference frames, they can be compared with each other. It is useful to work with a simplified setup with frames by a standard configuration.

Again, in special relativity, mass–energy is closely connected to momentum. Just as space and time are different aspects of a more comprehensive entity of space–time, mass–energy and momentum are merely different aspects of a unified, four-dimensional quantity called 'four-momentum'. In Poincare's conventionalist views, the essential criteria according to which one should select a Euclidean vs. non-Euclidean geometry would be economy and simplicity. Any realist would say that Einstein discovered space–time to be non-Euclidean, but any conventionalist would say that Einstein merely found it more convenient to use non-Euclidean geometry. The conventionalist would maintain that Einstein's analysis said nothing about what the geometry of space–time is.

Space–Time Distortion

Any object distorts the fabric of space–time, and the bigger it is, the greater the effect will be. According to the theoretical concepts, matter and energy distort space and time, curving them around themselves. Frame dragging theoretically occurs when the rotation of a large body twists nearby space and time. Gravity is the curvature of space and time. It is here that Einstein connected the dots to suggest that gravity is the warping of space and time. If space can be bent like, say, gravity, then space–time can also be bent. One may have seen space–time portrayed as a fabric, manipulated by energy. If space–time can be bent, it is theoretically proved that time can also be bent.

Given the ability to distort the space–time continuum, then a combination of space–time manipulation and distortion manipulation is also called 'continuum distortion' or 'space–time bending/warping'. Users can create distortions within the space–time continuum, the very fabric in which the world exists, allowing them to achieve a plethora of effects throughout space and time. In a relativistic Universe, time cannot be separated from the three dimensions of space. This is because the observed rate at which time passes depends on an object's velocity relative to the observer. Also, the strength of any gravitational field slows the passage of time for an object as seen by an observer outside the field.

In condensed matter, the vanishing of S-T D can induce a quantum phase transition with a strong resonant peak. However, we observe a remarkable similarity with some recent reports describing the S-T D found in micro ring resonators. Such a circuit, called the 'PANDA ring circuit', has already been discussed in the context of micro black holes where a set of three micro rings was employed to minimize the S-T D. By applying an electric field, or Kerr effect, S-T D can be created in the micro ring circuits. A signature of a black hole is the vanishing of S-T D, which is reflected in the observation of a strong resonant peak and also two satellite peaks. In the micro ring Kerr effect, the S-T D circuit represents a three-level system with spin states $ms = 0$, $+1$, and -1. In a spin-triplet superconductor,

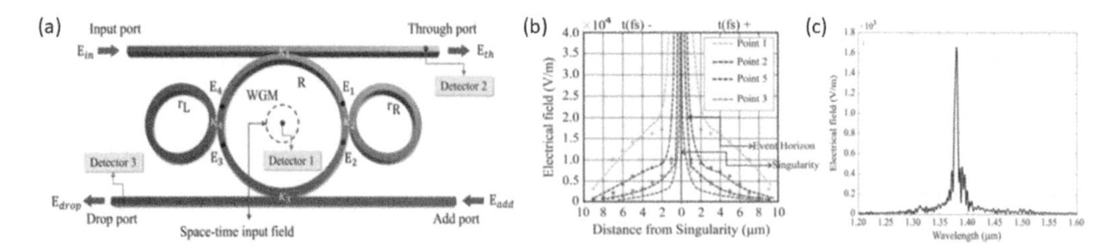

FIGURE 9.1
(a) PANDA micro ring resonator for controlling space–time distortion. An electrical circuit consisting of one central ring and two smaller rings on either sides of it can produce strong electric fields. (b) Variation of the electric field with the distance from singularity. (c) Wavelength shows strong resonance features.

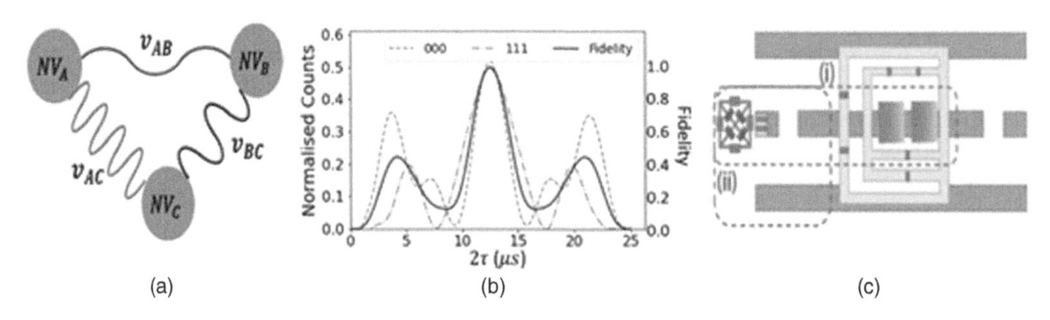

FIGURE 9.2
(a) Three NV centers are coupled together asymmetrically in a triangle formation and produces three peaks that can be tuned by the coupling. (b) Return probability of electrons plotted as a function of time shows three peaks [4]. (c) Simulation using superconducting flux qubit and two NV centers [5].

the spin states can be represented as $|\downarrow\downarrow>$, $|\uparrow\uparrow>$ and their superposition as ($|\uparrow\downarrow> + |\downarrow\uparrow>$), which correspond to the side peaks and the central peak, respectively. This effect can also be found in the spin-orbit coupling interaction, which causes the signature of spin-triplet superconductivity in some unconventional superconductors and which can be useful in developing a new range of quantum devices.

These features can be observed in a spin-triplet superconductor as a strong zero-bias conductance peak with two satellite peaks. Similar results can be obtained from three qubits, which can be a set of either three or two nitrogen vacancy (NV) centers combined with a superconducting flux qubit.

In this way, an artificial (holographic) vision can be developed in microelectronic devices. By designing proper circuits (micro resonator rings), not only stereoscopic vision but also three-dimensional (3D) vision can be achieved. This is particularly important when a third-eye-like structure producing 3D imaging is presented. The 3D printed sensors can be applied for the recognition of human body gestures. It is claimed that these micro ring sensors can form a human-like stereo system. For the eyes, it will give a 3D vision. This discussion will be added to Chapter 10 on quantum devices and quantum consciousness; hence mind–matter interactions are explained.

We further discuss the complex nature of space–time distortion (in distorted vortices) through a holographic model, i.e., the Sachdev–Ye–Kitaev (SYK) model, which can explain quantum phase transition (see on page 139). The SYK model with interactions has great applications in describing the quantum many-body problem, which involves strong localization, tunneling between chaotic quantum dots, and many other non-Fermi liquid behavior and which can be demonstrated in the coupled SYK system.

At present, we note that the space–time distortion in the SYK model has not been developed yet. Also, experimental demonstrations of the features of a pure SYK model having no quasiparticle interaction have been attempted; however, the realization of the SYK model in a solid-state system has appeared to be difficult. There are attempts to understand the complexity of the SYK model and black holes through quantum simulations based on superconducting qubits, but more work is needed in the context of S-T D. Therefore, we attempt to unify the concept of disorder through the S-T D. Using a set of superconducting qubits in a loop configuration, we demonstrate the presence of the traversable wormhole in the low-temperature limit as described by the Maldacena and Qi model and its dual, i.e., two copies of Sachdev–Ye–Kitaev (SYK) model, connected by a tunnel path. We see a strong central peak with two satellite peaks in the return probability of electrons (see on page 142).

Black Hole

A black hole, a cosmic body of extremely intense gravity, is a region of space–time where gravity is so strong that no particles or even electromagnetic radiation such as light can escape from it. Black holes are some of the strangest and most fascinating objects in outer space; they are also extremely dense. The theory of general relativity predicts that a sufficiently compact mass can deform space–time to form a black hole. A black hole is a tremendous amount of matter crammed into a very small, in fact, zero amount of space. The result is a powerful gravitational pull, from which not even light can escape and therefore no information or insight can be gathered as to what life is like inside. A black hole can be formed by the death of a massive star. When such a star has exhausted the internal thermonuclear fuels in its core at the end of its life, the core becomes unstable and gravitationally collapses inward upon itself, and the outer layer of the star is blown away. The crushing

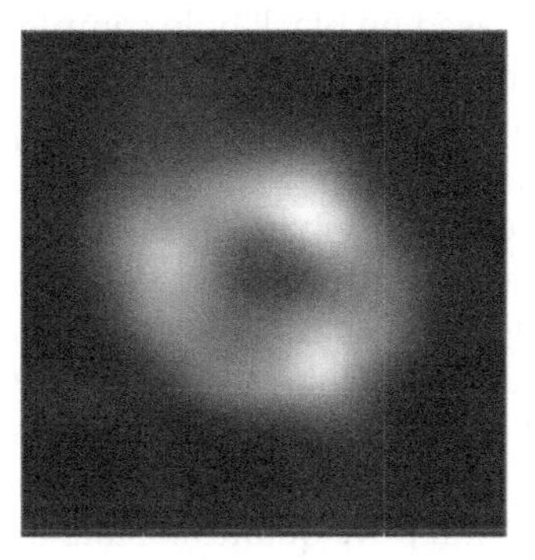

FIGURE 9.3
First image of Sagittarius A*, the supermassive black hole at the center of the Milky Way, as captured by the Event Horizon Telescope. (Credit: EHT Collaboration).

weight of the constituent matter falling in from all sides compresses the dying star to a point of zero volume and infinite density called 'singularity'. The singularity at the center of a black hole is a place where matter is compressed down to an infinitely tiny point and all conceptions of time and space completely break down: It does not exist. Something has to replace the singularity, but we are not exactly sure about that.

Black holes can be very tiny but have the mass of a large mountain. Mass is the amount of matter or 'stuff' in an object. Another kind of black hole is called a 'stellar black hole', whose mass can be up to 20 times more than the mass of the Sun. There may be many, many stellar-mass black holes in the galaxy of the Earth (Milky Way). Stellar black holes are formed when the center of a very massive star collapses in upon itself. This collapse also causes a supernova, or an exploding star that blasts part of the star into space. Scientists think that supermassive black holes are formed at the same time as the galaxy they are in. These supermassive black holes have masses that are more than 1 million Suns together. Scientists have found proof that every large galaxy contains a supermassive black hole at its center. The supermassive black hole at the center of the Milky Way is called 'Sagittarius A′. It has a mass equal to about 4 million Suns and would fit inside a very large ball that could hold a few million Earths. Scientists think that the smallest black holes were formed when the Universe began.

Stellar black holes are formed when the center of a very big star falls in upon itself or collapses. When this happens, it causes a supernova. A supernova is an exploding star that blasts part of itself into space. Most black holes form from the remnants of a large star that dies in a supernova explosion (smaller stars become dense neutron stars, which are not massive enough to trap light). Scientists think that supermassive black holes were made at the same time as the galaxy they were in.

A black hole is created through the packing of matter into a very small area. This results in a strong gravitational field, that nothing, not even light can escape. The idea of an object in space being so massive and dense that light could not escape it has been around for centuries. Most famously, black holes were predicted by Einstein's General Theory of Relativity, which clearly showed that, when a massive star dies, it leaves behind a small, dense, remnant core. If the core's mass is more than about three times the mass of the Sun, the equations show that the force of gravity overwhelms all other forces and produces a black hole.

It is suggested that a black hole does not have any loss of information. According to Hooft and Susskind, all of the information contained in a volume of space can be represented by a theory that is based on black hole complementarity: Information doesn't travel faster than light. The outside observer sees it come out, the infalling observer sees it inside, and they can't compare notes—a new relativity principle. The ultimate limit to the storage of information is that if you try to pack more and more information onto your hard drive, then eventually the drive will collapse into a black hole. Stephen Hawking and Kip Thorne firmly believe that information swallowed by a black hole is forever hidden from the outside Universe and can never be revealed even as the black hole evaporates and completely disappears. John Reskill firmly believes that a mechanism for the information to be released by the evaporating black hole must and will be found in the correct theory of quantum gravity. Therefore Preskill offers and Hawking/Thorne accept a wager that: When an initial pure quantum state undergoes gravitational collapse to form a black hole, the final state at the end of black hole evaporation will always be a pure quantum state.

A Maximally Chaotic System

Quantum systems allow one to simulate a multitude of effects that have a quantum nature and that thus cannot be fully realized and explored via conventional classical computation avenues. Several different models have been simulated with this ranging from spin models, quantum phase transitions, superconductivity, topological order, and several other areas in condensed matter physics. Among all the models, the Sachdev–Ye–Kitaev (SYK) model with interactions has great applications in describing the quantum many-body problem, which involves strong localization, tunneling between chaotic quantum dots, and much other non-Fermi liquid (NFL) behavior that can be demonstrated by the coupled SYK system. The SYK model is a model that describes a non-Fermi liquid with maximally chaotic behavior. It can connect non-Fermi systems with theoretical concepts such as quantum gravity through a holographic principle. The thermal state of this system is called the Sachdev–Ye state, which is a non-Fermi liquid whose entropy is nonzero. At low temperatures, this model is shown to be maximally chaotic. This state is dual to charged black holes with AdS_2 horizons. The SYK model has an advantage over other holographic models because the particles it is made up of are not relativistic.

The SYK model consists of *N* mode Majorana fermions with two body-hopping between particles. This causes the fermions to be entangled with one another, and thus the system can be described as being maximally entangled. A Majorana fermion is an exotic particle that is its antiparticle. They are fermions, which means that they are elementary particles with a half-integer spin and are governed by non-Abelian statistics. Bound Majorana fermions can appear as quasiparticle excitations in condensed matter physics. The SYK model is solvable in the large *N* limit. The principle of holography claims that black holes and quantum gravity are equivalent to nongravitational theories in different space–time dimensions. If the nongravitational system is defined on its boundary with dimension *d*, then the space–time bulk of the quantum gravitational system will be (d +1) dimensional. Maldacena's duality constructs quantum gravity in a very special box, known as 'anti-de Sitter space', that gauges the theory on surface and gravity in the interior. The quantum theory of a black hole in a 3+1-dimensional negatively curved AdS_2 Universe is holographically represented by a CFT (the theory of a quantum critical point) in 2+1 dimensions. The SYK model in a strongly correlated system is holographic due to the extremal black hole in AdS_2 space. The holographic principle is very different from previous physical laws.

Holography

Let's understand some basics of holography.

Holography is the science and practice of making holograms. A hologram, also known as a holograph (whole description or whole picture), is a recording of an interference pattern that uses diffraction to produce a 3D light field, resulting in an image that still has the depth, parallax, and other properties of the original scene. A hologram is a photographic recording of a light field rather than an image formed by a lens. Holograms are now entirely computer-generated in order to show objects and scenes that do not exist. The holographic medium—for example, the object produced by a holographic process, which may be referred to as a hologram—is usually unintelligible when viewed under diffused ambient light. It is an encoding of the light field as an interference pattern of the variations in the opacity, density, or surface profile of a photographic medium. When suitably lit, the interference pattern diffracts the light into an accurate reproduction of the original light field, and the objects in it exhibit visual depth cues such as parallax and perspective, that change realistically with the different angles of viewing—that is, the view of the image from different angles, representing the subjects, viewed from similar angles. In this sense, holograms do not have just the illusion of depth but are truly three-dimensional images.

For a better understanding of the process, it is necessary to understand interference and diffraction. Interference occurs when one or more wavefronts are superimposed. Diffraction occurs when a wavefront encounters an object. The process of producing a holographic reconstruction is explained here purely in terms of interference and diffraction. It is somewhat simplified but is accurate enough to give an understanding of how the holographic process works. Holography is a technique that enables a light field (which is generally the result of the light source scattered off objects) to be recorded and later reconstructed when the original light field is no longer present due to the absence of the original objects.

The term 'Maya' described earlier as Great Illusion is closely related to holography. In this description, two particles rotate in opposite directions at the same time. This is a duality that can create a three- or multi-dimensional space through the emission or absorption of energy like heat emission, which flows back to the system. This backflow is precisely the nature of Shakti, who maintains a feedback loop. The surface of the event horizon is flat or two-dimensional and becomes three-dimensional at the black hole. As space merges with the black hole, some energy comes out as a result of feedback (backflow). This is like Tamas or the Great Illusion (Mahamaya). Since the dimension increase is assisted by the breakdown of space, attraction to the center increases. Finally, it breaks down with the release of energy (Fire). Folding the space increases the mass, which is the division of itself or a zero (vacuum). This folding or division is watched by Vishnu through his dream, which goes to a singular point: Brahma. This is like the Sun that watches us throughout time in association with the night and three characteristics, i.e., dawn, noon, and dusk. Everybody is watching the progress of life on Earth. If anything blocks the vision, it will be destroyed eventually through this process.

Quantum Chaos and the SYK Model

The SYK model exhibits connections to several areas of interest in physics such as quantum chaos, holographic correspondence in black holes, and quantum information scrambling, among others. It is a quantum mechanical model of N strongly interacting Majorana fermions with all-to-all random couplings. An approximation of the SYK model can be obtained in this system with the superconducting quantum boxes that are coupled to create a chain, whose interaction can be used to emulate this effect. In a chaotic system, there is always a competition between the direct hopping (tunneling) and random hopping processes, which are attributed to the chemical potential and exchange coupling terms of the SYK Hamiltonian. The competition can be controlled by the system temperature as well as through the connectivity between qubits, e.g., the distance between the atoms and the geometrical arrangements. The dispersive SYK model has been proposed by introducing a constant hopping term to the SYK Hamiltonian, including a random hopping term. The picture of random vs. direct hopping or tunneling can be simulated by the IBM QE with a linear chain and square lattice configurations of SC qubits, respectively.

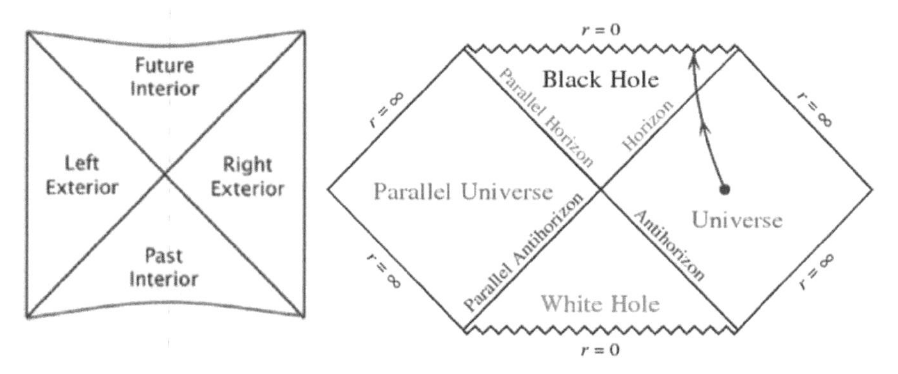

FIGURE 9.4
Penrose diagram of the complete, analytically extended Schwarzschild geometry. The diagram looks like a square having four areas. The top and the bottom ones represent the future and the past interiors, respectively. In the other diagram they represent a Black hole and a White hole. Whereas the side areas represent the left and the right Exteriors, or a parallel universe and a universe, respectively.

Wormhole

A wormhole or Einstein–Rosen bridge or Einstein–Rosen wormhole is a speculative structure linking disparate points in space–time and is based on a special solution of the Einstein field equations. A wormhole can be visualized as a tunnel with two ends at separate points in space–time (i.e., different locations, different points in time, or both). For a simplified notion of a wormhole, space can be visualized on a two-dimensional surface. In this case, a wormhole would appear as a hole in that surface leading to a three-dimensional tube (the inside surface of a cylinder), then reemerge at another location on the two-dimensional surface with a hole similar to the entrance. An actual wormhole would be analogous to this but with the spatial dimensions raised by one. For example, instead of circular holes on a two-dimensional plane, the entry and exit points could be visualized as spherical holes in three-dimensional space leading into a four-dimensional tube similar to a spherinder. Wormholes are consistent with the general theory of relativity, but whether wormholes exist remains to be seen. Many scientists postulate that wormholes are merely projections of a fourth spatial dimension, analogous to how a two-dimensional being could experience only part of a three-dimensional object. Theoretically, a wormhole might connect extremely long distances such as a billion light-years, short distances such as a few meters or different points in time, or even different universes.

If traversable wormholes exist, they might allow time travel. A proposed time travel machine using a traversable wormhole might work hypothetically. One end of the wormhole is accelerated to some significant fraction of the speed of light, perhaps with some advanced propulsion system, and then is brought back to the point of origin. Alternatively, another way is to take an entrance of the wormhole and move it to within the gravitational field of an object that has higher gravity than the other entrance and then return it to a position near the other entrance. For both these methods, time dilation causes the end of the wormhole that has been moved to have aged less than the stationary end as seen by any external observer; however, time connects differently through the wormhole than outside it, so that synchronized clocks at either end of the wormhole will always remain synchronized as seen by observer passing through the wormhole, no matter how the two ends move around.

Black Holes and Worm Holes

Recently, Maldacena and Qi proposed a quantum model of a wormhole formed by direct hopping or tunneling with a large probability between two copies of the SYK model at low temperatures. The holographic duality between the SYK model and black holes with AdS_2 dilution gravity can be made stronger if the duality between a coupled pair of SYK model and the MQ model is firmly established. The coupling can be established through tunneling or hopping between two chaotic quantum dots and also a geometrical phase that can be introduced in the qubit configuration (series and parallel). However, the full potential of the SYK model has not been explored yet, particularly the connectivity of two or more copies of the SYK model through quantum simulation. This study can be extended to traversable wormholes and the Hawking–Page transition in a coupled complex SYK model.

Experimental demonstrations of the features of the pure SYK model having no quasi-particle interaction have been attempted; however, the realization of the SYK model in a solid-state system has appeared to be difficult. So far, features of non-Fermi liquids are simulated by the SYK model using NMR, photonics, and cold atoms, but the full interaction picture in coupled SYK model relating to the black hole model has not been given. There are attempts to understand the complexity of the SYK model through quantum

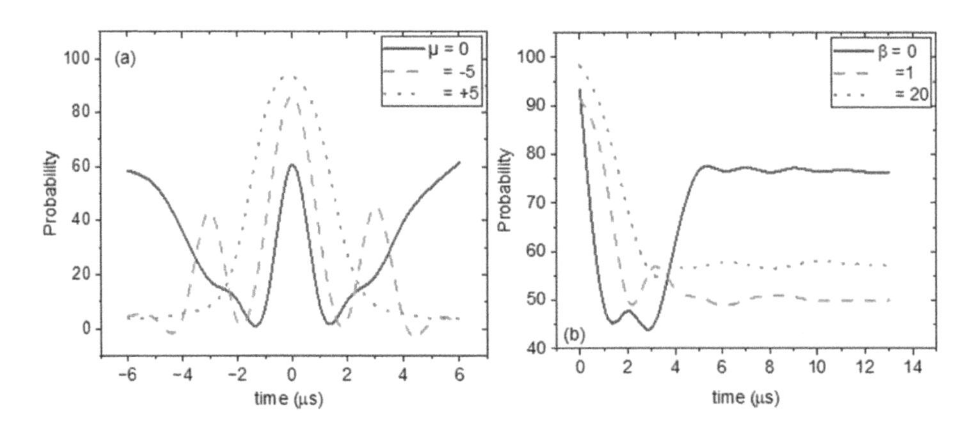

FIGURE 9.5
SYK combined with weak localization correction (for the linear chain configuration). (a) The probability distribution shows a central peak and two side peaks which can be tuned by the chemical potentials μ. The distribution is shown for the inverse temperature $\beta = 20$ and the varying values of μ (0, −5, +5) by solid, dashed, and dotted lines, respectively. The intensity of the central peak increases from −-5 to 0 and further increases for +5. For a double-path system, prominent oscillations can be seen in addition to satellite peaks; this explains the experiment data particularly at low temperatures and for $\mu = +5$. (b) The OTOC curves are plotted for a double path system and for $\beta = 0$, 1, and 20 representing solid, dashed, and dotted lines, respectively. The probability decreases for very small values of temperature with time however, rises for high temperatures. These results explain the nature of a traversable wormhole in a double SYK system (unlike in the single SYK system).

simulations based on superconducting qubits. Here we show IBM QE superconducting qubits for the first time employed for the experimental demonstration of a coupled SYK model and its dual to the wormhole where quantum interference correction has been added. We show how the boson correlation function changes with time from a pure SYK (a chaotic NFL state) to a nonchaotic ground state in the presence of coherent backscattering. Other interesting features, particularly the scrambling of quantum information in black holes, can establish the holographic duality with the SYK model through out-of-time-order (OTOC) measurements.

The Lyapunov exponent that defines the classical chaotic motion can be determined from the frequency and temperature dependence of the WL correction. The WL correction of antidots can explain disordered vortices, which is equivalent to the maximally connected chaotic quantum dots connected by coherent backscattering paths. Therefore, we added the quantum interference effect or WL correction to the quantum circuits, which can also be dependent on the temperature and the chemical potential of the system. More importantly, the coherent backscattering or the WL can show an 'echo' effect that can demonstrate the transition from a chaotic NFL to a nonchaotic pair condensate through its oscillatory nature. Such an echo has been recently observed from black holes. The dispersive nature of quantum chaos and many-body localization in the SYK model can be explained by the OTOC measurements where both WL corrections and spin-flipping can be added to a quantum circuit. Quantum tunneling and the oscillatory feature of magnetoresistance in a chaotic quantum dot structure have been studied in the presence of weakly localized transport, which we simulate in this work. Instead of scrambling quantum information, revival oscillations can be seen in connected black holes that are supported by the MQ model and presented in this article. The quantum chaos in two copies of the SYK model, coupled through a tunnel barrier in the presence of coherent backscattering, can act as a dual of Jackiw–Teitelboim (J-T) gravity. Hence the Hawking–Page-type phase transition from the gapped state at low temperatures in superfluids to a gapless spin-triplet state has been simulated. In the gapless superconductor, this kind of transition can be seen through the zero-bias conductance peak with satellite peaks and an oscillatory feature of magnetic field-dependent resistance, which can be explained by the SYK model and simulated here.

We calculate OTOC for two copies of SYK connected by a wormhole. In the cold atom system, OTOC measurements were performed through the laser excitation of an atom or a dimer having a double-well potential with a molecular state. The superposition of the atomic and the molecular state can form two double-well potentials in the superposition state. The overall process can be explained by the interaction of a control qubit and a target qubit. According to Einstein's gravity theory, in a black hole, there is an exponential decay of OTOC associated with the scrambling of information. At high temperatures, the scrambling of information increases, and the OTOC drops drastically to zero with the increase of time. At low temperatures, the tunneling rate increases, and the OTOC saturates or has a slow decay. The tunneling is mediated by the chemical potential of the SYK system.

The interesting results of a revival of the OTOC (Figure 9.5) for the coupled SYK system raise the question of how quantum teleportation takes place. The idea of the traversable wormhole is developed based on the thermofield double state (TFD) consisting of copies of two identical quantum mechanical systems. The gravity dual of the TFD state is a two-sided black hole with two asymptotic boundaries showing the recovery of information from Hawking radiation. It is said that the process of recovery of information generates a new space–time that is connected to the original black hole through the interior. In the paper of Maldacena et al., the variation of a correlator was studied in the absence and in the presence of backreaction that accounts for the teleportation between two black holes.

In a superconducting or condensed state, we see quantum vortex structures that give reentrant superconductivity. It is said that the process of recovery of information generates a new space–time that is connected to the original black hole through the interior. In the paper of Maldacena et al., the variation of a correlator was studied in the absence and in the presence of backreaction, which accounts for the teleportation between two black holes. The results are very similar to a gapped superconductor and gapless or a spin-triplet superconductor, respectively. The structure of a black hole can be compared to a quantum vortex that can be simulated through a two-path WL system. Here we present a very simple picture using a two-path weak localization system that includes coherent backpropagation, and hence they look similar to those describing traversable wormholes. See references on page 177.

Summary

The SYK model was developed based on the maximally chaotic nature of the lattice, which may represent a supermassive as well as a normal structure. The features of a black hole were shown as a strong zero-bias conductance peak with two satellites that forms a Trinity and that is also a feature of spin-orbit interaction of a spin-triplet superconductor. Finally, we combined two black holes and found the features of a traversable wormhole. We showed how teleportation works between two black holes or two vortices. A complete model of the Universe by combining two vortices was developed, which is self-sustained through a feedback process. I find that this description of the Universe is very consistent with the one given in ancient scriptures as described in the preamble, particularly a two-egg model. In Chapter 10, we see how a feedback loop leads to artificial intelligence.

10

Quantum Simulation, Artificial Intelligence, and Vision

In Chapter 9, we learned that creation is not free from disturbances and noise, which can be suppressed by recycling (a feedback process). In this last chapter, I shall bring an ancient story of rotating the Universe that is key to the 'creation' process.

- Philosophy of Artificial Intelligence
- Weak Localization
- Process Color and Geometric Phase
- Emulating Spin-Orbit Interaction
- Holonomic Quantum Computation
- Optimization and Image Processing
- MCQI for Artificial Vision

Samudra Manthana: The Churning of the Ocean

The mythological *Samudra-Manthana* explains the origin of Amrita, the nectar of immortality obtained from the churning of the ocean. Similar stories can be found in various Indo-European myths and even in the European medieval legend of the Holy Grail. In this Old Norse poem, a sacred mead is prepared by cooperating with the gods and the giants (who might respectively correspond to Devas and Asuras in Indian mythology), with the gods ultimately winning the drink.

Long ago, there was a battle between gods and demons. The gods were defeated, and the demons gained control over the Universe and heaven. The gods sought Vishnu's help, who advised them to treat the demons diplomatically. The gods allied with the demons to jointly churn the ocean for the nectar of immortality and to share it among themselves. However, Vishnu told the gods that He would arrange for them alone to obtain the nectar. The churning of the Ocean of Milk was an extensive process: Mount Mandara was used as the churning rod, and Vasuki (king of the serpent) who abides on Shiva's neck, became the churning rope.

The *Samudra-Manthana* process released several things from the Ocean of Milk. One of them was the lethal poison. However, in some other variations of the story, it is mentioned that the poison escaped from the mouth of the serpent king as the demons and gods churned. This terrified the gods and the demons because the poison was so powerful that

DOI: 10.1201/9781003304814-11

FIGURE 10.1
A model shows creation process through the battle between gods and demons who are churning the ocean pulling a large snake in opposite directions.

it could destroy all of creation. In the variation, Vishnu knew that Vasuki would vomit the poisonous flames when twisted and pulled and therefore advised the gods to hold the head end of the snake while the demons held the tail end. The demons were enraged by this as the lower part of an animal is fully impure or less pure than the part that contains the head. They insisted on holding the head side of the snake. Vishnu had an inkling that His reverse psychology would work. The demons demanded to hold the head of the snake, while the gods, taking advice from Vishnu, agreed to hold its tail. When the mountain was placed in the ocean, it began to sink. Vishnu in the form of a turtle came to their rescue and supported the mountain on His shell.

The demons were poisoned by fumes emitted by Vasuki. Despite this, the gods and the demons pulled back and forth on the snake's body alternately, causing the mountain to rotate, which in turn churned the ocean. The gods then approached Shiva for protection. Shiva consumed the poison to protect the three worlds, which in the process gave a blue hue to His throat. In some versions, as Shiva drank the poison, He suffered intense pain but could not die, as seen by Goddess Parvati (Durga). She immediately placed a hand on His throat, stopping the poison to flow any further, and Her illusion (Maya) stopped it forever. As a result, His throat turned blue.

In this chapter, we show how rotation qubits can produce a geometric phase that can explain all the quantum phenomena described earlier.

In the previous chapters, we have seen interactions between spin one and spin half particles or the competition between magnetic and electric fields. The duality is destroyed by establishing an equilibrium or a trinity. The purpose of this process is to extend our vision through the addition of an extra dimension. Gods, who cannot be manifested, can create their images and produce a hologram and be visible indirectly. The manifestation is called 'imaging' or 'producing a representative' of the god, which can be treated as the son of god or the consort of god. For example, Vishnu created Brahma and also Maya Shakti from his body before rising from the limitless bed. Three prime gods, namely Brahma,

Vishnu, and Shiva, created Goddess Durga by combining their visions. Also, Durga created Goddess Kali from her forehead (or mouth). Maya or an illusion can be created by quantum tunneling through the image creation of an invisible object separated by a barrier. In this way, gods created their projections without their movement, which can be extremely useful in fighting against any resistance or demons. In science or mathematics, a hologram can be created by constructing a solid angle, i.e., a triad that is explained in the spin-orbit interaction (in Chapters 7 and 8). In this way, a geometric phase can be constructed by rotating the qubits. All these quantum processes can be simulated by a set of qubits that can be arranged as a quantum processor or a quantum computer that is commercially available today.

Today, superconducting qubits can be considered ideal building blocks for the quantum computers of the future. The quantum oscillations observed in experiments so far have amplitudes that are at best about half of what they should be. We cannot tell whether this is due to certain problems with the qubits or inadequate measurements, and we have to work hard on improving both aspects. We also need to increase decoherence times, in particular by using improved fabrication techniques to reduce the number of defects in the tunnel barrier. This is not going to be easy or happen quickly. Improved fabrication techniques are also needed to improve the yield and quality of samples, which will speed up research. There is also a need for better ways to couple qubits with one another and also with the outside world and for new ways to switch this coupling on and off. We can expect progress in this area in the future, particularly in noise reduction to produce artificial intelligence.

Artificial Intelligence

Artificial intelligence (AI) is the wide-ranging branch of computer science, concerned with building smart machines, capable of performing tasks that typically require human intelligence. It is the endeavor to replicate or simulate human intelligence within machines. Thus, artificial intelligence (AI) refers to the simulation of human intelligence in machines that are programmed to think like humans and mimic their actions. The term may also be applied to any machine that exhibits traits associated with the human mind such as learning and solving problems.

The common three philosophical questions that arise regarding artificial intelligence are (1) whether artificial intelligence is possible, whether a machine can solve any problem that a human being can solve using intelligence, or are there hard limits to what a machine can accomplish; (2) whether intelligent machines are dangerous and how humans can ensure that machines behave ethically and that they are used ethically; (3) whether a machine can have a mind, consciousness, and mental state in the same sense that human beings do. Can a machine be sentient and thus deserve certain rights, and can a machine intentionally cause harm?

The core aim of the traditional field of artificial intelligence (AI) is to develop classical computer systems that can think like a human brain and reinforce that with classical computers' high speed of execution and ability to access huge memory for storing and manipulating data. This can be achieved through the choice of appropriate quantum materials as well as device fabrication using advanced nanotechnology. Neural networks and closed feedback loops in superconducting circuit (SC) technologies have already been considered building blocks for AI and quantum machine learning. The performance of standard SC can be greatly enhanced if spin qubits (SQ) are introduced in the system where the SQ can work as extra feedback loops. One of the best SQ is diamond NV centers, which can also perform the holonomic quantum operation and compared to spin-orbit coupling effects (explained later in the chapter). In a hybrid circuit consisting of a diamond flux qubit and NV centers, the squeezed cat states are used to increase the fidelity of QC. This approach can be used for the development of artificial vision.

Philosophy of Artificial Intelligence (Vision): Realizing the 'Third Eye'?

While searching a broad area of artificial intelligence, the relatively new work I find is on artificial vision. Did *Vishwarup Darshan*, or the vision of the Universe in the Hindu scripture, *Bhagavad Gita*, and most importantly in Sankhya philosophy, have something to say about this? The static Purusha is the source of knowledge (or heat) that always radiates light. Prakriti collects energy from Purusha by creating circles (voids or Sunya) around Him. Circles of different sizes are collected in a very small volume as Mahat, which creates Buddhi. At the critical energy or size, this gives birth to more circles in a very symmetric manner. The addition of three Gunas produces circles of different shapes and sizes. These are called 'atoms' consisting of several orbits in a small invisible body. When many atoms meet together, we can see them physically as fire, water, or matter, or we can experience the effect like air and also the sky or free spaces (voids). However, no experimental proof supports philosophy since Purusha and Mula Prakriti cannot be visualized directly.

We extend the knowledge gained from our daily work such as cooking food to computers operated by artificial intelligence. Machine learning algorithms are classified into three categories: supervised, unsupervised, and reinforcement learning. They are based on a series of classified data, unclassified data, and reward processes that can be compared to *Tamas*, or a large amount of known data; *Satwa*, which are independent of data (based on correlations in available data); and finally the *Rajas*, or exchange interaction with the environment based on optimization or deficiency. The third one, i.e., reinforcement learning, is based on feedback loops. These three types of learning are connected and dependent on one another and also form an equilibrium. According to Sankhya philosophy, Prakriti

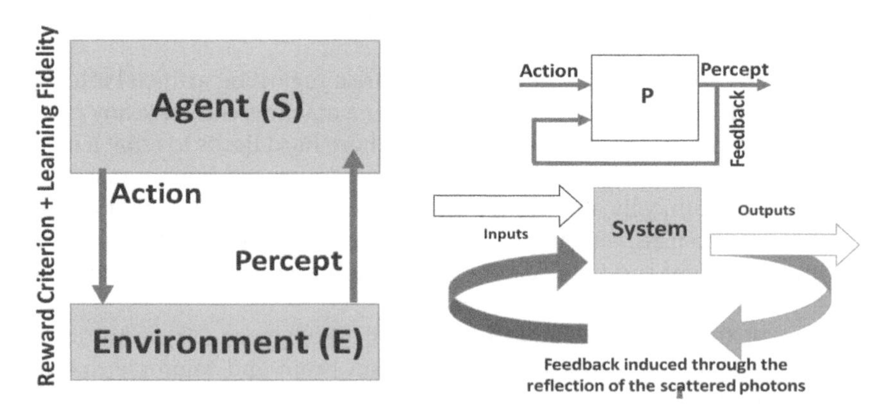

FIGURE 10.2
A flowchart shows the interactions between an agent and the environment as a feedback loop.

gains consciousness through interaction. She converts the energy of Purusha through a time-dependent function (or creates time), which is Shakti. Then she creates 'space'. She contains all information extracted from Purusha through duality. This process is the creation of an atom that has a defined space. Since there is wave–particle duality, a particle (mass) model can also be constructed. So energy is converted into a mass that is a form of consciousness. This space contains a dual nature that is disorder or chaos. This felled space, Mahat, should not be static but can be rotated to extract knowledge, or Buddhi. So break it up into three Gunas or a triplet state. Two sidelines are employed to satisfy the machines by rewarding them as in reinforcement learning. The middle one is a pure feedback loop. It is like a motor that can be used with a gear system. The motion of a motor is transferred via the gears to another point of action. Sidelines can also form feedback loops to support the rotation of the central feedback loop rotating in a normal plane. This looks like a moving mass or center of mass from one point to another. Without this process, information cannot be propagated. One feedback loop represents one orbit, but for a complex process, we need a multifeedback loop; hence multiple orbits. Energy transfer can be achieved through multijump processes between orbits. This is like a neural network. Circular orbits can be mapped into a multilayered system consisting of multiconnectivity. So space or atom is split into four branches, i.e., knowledge, work, sense, and elements, and is connected by some links or mind, i.e., Rajas. They are four pillars of Shakti that sit on the body of Shiva and also on a plane. This describes five Tanmatra of Sankhya philosophy. The space Mahat prohibits the return of particles or feedback. Although there is not much advancement in the science of the brain that produces thought, we believe that this is the way the human brain or nervous system should work.

If we understand the cause (or reason), then we can find the effect (or the answer). The cause is associated with elementary units. Finding the characteristics of the elementary units, including size, shape, speed, color, and temperature, will be useful for constructing a large structure. For example, if we know circles, we can add them up and construct an object. However, we have to know what we are going to construct and arrange or connect the circles accordingly. Then if you know circles and lines, you can draw any picture. You can also resolve a complex pattern into circles and lines orientated at different angles. This works for artificial vision when an image is not understood. Computer vision breaks it into small known objects, i.e., circles, triangles, etc. A computer remembers these known patterns through some mathematical languages. Then an AI-operated camera sees a real object as a combination of circles etc. as a known pattern and compares it with its stored memory. This comparison identifies the objects and then reconstructs an image.

How do we connect the objects? Simply, we like to see symmetry in nature. Symmetry is formed by the state of equilibrium. We break the symmetry and create a distorted picture. However, the symmetry is regained if we adjust three objects that oppose one another, such as red, green, and blue colors. A more complex image or shape should be analyzed with the principle that can be inspired by nature. The Sun, the source of power to us heats up the planet Earth every day and cools it down at night. If the Earth does not rotate on its axis, then the part that faces the Sun will be too hot for survival and the creation of life. If it has no source of power, then the Earth will be floating in the cold Universe, and life cycles will not be started. So heat and rotation are connected. Philosophically, the Sun is the prime god that gives power to the Earth, and in return, the Earth obeys the Sun's gravity by rotating about its axis and orbiting the Sun. In this way, nature is established. Interaction between the Sun and Earth creates the energy or transfer of energy from the infinite source: the Sun to the Earth. It creates all forms of energy on our planet. This energy cannot be stored forever and must be spent on actions or work. That means that we need forces that are the cause of the actions. This force is called 'Buddhi', or intelligence, which comes from the Sun to Earth through the transfer of heat in the form of a cycle. This cycle is used in computer programming or machine learning. The program (or software or algorithm) drives the hardware or power of a computer. A complex program needs more power from the computer. The supplied current or heat to the computer is converted into the operation of the program. AI can be produced if the machine can produce its heat, which is possible through a feedback loop. You also need an intermediate step to distribute the heat properly.

But this is merely a mechanical process without a proper command or direction. You need a third object who can direct this process, an observer, whose wish is transformed into work. This is the mind, or Maya, which resembles the phase of the Moon. The purpose of the mind is to create diversity in nature which can be of different shapes, colors, sounds, heat (odor), and sizes (sky/volume). Mind binds all elements and unifies them to one point, which is Prakriti, and eventually to Purusha. It comes from the mind of the Sun, which likes to create its replica, the Moon, at night. It rotates and revolves and wishes to create an object with a very similar nature. Moon fulfills Sun's wish, which is an artificial sun to us, especially when we need light at night. Also, it has a magnetic field and wants to create magnetic fields associated with the Earth and the Moon.

So we understand that an artificial object is created from an image or dividing one into two. The second one possesses a very similar or complementary nature to the original one, just as we want to see something boiling in the pot when we apply heat. We are happy to see it is forming by bubbles in the liquid. We stop the process when we see it is going to be no longer necessary to heat the liquid. So the fundamental thing of creation is to divide the object that allows interactions, make more numbers, and finally assemble all parts. The mind can do this job efficiently. Mind is something having a mass that can be achieved through an attractive force that can break and make things. This has to be added to a machine that can think of a process.

Division comes from sufficient (Bhav) and deficiency (Abhavab). Mostly, deficiency creates intelligence. Let's make AI with this property. A machine cannot see something that it would like to see. If you ask a machine not to work, it will try to work, which is the representation of the intelligence of sense. My mind does not want to see coffee as cold; it wants to see steam coming out of the cup and the milk *not* static but swirling. I increase the heat until I see the milk is moving or rotating and will be satisfied. Create a circular path or void in the memory of an AI-powered computer, and you create an imaginary world. The lack of a particular color will produce artificial colors that the human brain cannot make. As a result, the disordered pattern of a machine will be different from that of a human who has a true sense of color. AI will try to regulate the colors with a different logic that is *absent* in the real world.

Deficiency or absence in the measure of Buddhi corresponds to how much space you open up during a process. This is related to the moment or angular momentum of particles moving in a large void of energy. If we shrink the size of the void, then movement decreases; however, since the angular momentum is conserved, we see things are moving at a faster rate. By creating a void or space, we can control nature, including shape or color or even heat. This process can be repeated by filling up space as well, which can change the appearance of the Universe. This theory works in a pot that boils liquids and even in a coffee mug. We can restrict the amount of milk added to the coffee to see nice or fine spirals as it goes toward the center of the mug, and the circle contracts. Adding a large amount of milk will spoil the fine picture. By regulating the heat in the pot, this spiral design can be controlled. An AI-powered computer can easily adopt this logic which sees things in the darkness or leaves them in silence or analyzes data that is buried in the noise. It will create a situation of a darker state or a noisier environment that is worse than an acceptable state and make an entangled state a good being (more light or sound). An interplay between dark and bright states will create a new image. It works just like the phases of the Moon. The lack of light on the surface of the Moon creates a nice-looking crescent.

Shapes can also be created from constriction. By putting constriction or by squeezing some part of the day on a wheel, a nice shape can be produced. By squeezing, we increase the frequency of the wave, and the wavelength is reduced. So by opening and closing the gate, we can adjust the depth of the field. This can be explained as a soft and sharp focus of light in an optical system. In instruments such as musical pipes, this creates high pitch or frequency when the mouthpiece is nearly closed and confined to an open pipe. If the pipe is open, then other effects like resonance can be produced. Ultimately, sharp focus will collect more information about the object. In a squeezed light, uncertainty of finding a particle is greatly reduced, which reduces noise levels. Mathematically focusing on light to create an image is done through a Fourier transformation of light waves. A lens is a Fourier transformer that converts three-dimensional objects spaced with distances into a two-dimensional plane where all objects can be seen at one time. A Fourier transformer converts invisible objects into sensible things like the movement of a speaker into audible frequencies. By squeezing wave functions, we can transform invisible to visible, i.e., from emitter to receiver information transmits. Work is converted into knowledge (reaction) in this way.

All five functions are done by the five organs—the eye, ear, nose, tongue, and skin—which can be equated with the five senses. This is the significance of duality, which is possible due to Fourier transformation or focusing. Similarly, fire, water, air, earth, and sky can be connected to sound, touch, smell, form, and taste. These five senses are found in an elementary biological system, even in a plant, but they cannot be implemented in an AI system available today; however, we have to think about how to mimic the natural process. So from a confined space, the elements can get separated, which can be observed distinctly. Our brain can differentiate among all five senses simultaneously. In quantum mechanics, all quantum numbers can be separated due to confinement into a very small volume. Similarly, all forces can be separated from each other in an extremely confined space like a black hole, the most intelligent object in the Universe. A black hole is a Maha Sunya, which absorbs everything including light, whereas the Universe is holographic. Plants can absorb light and convert it into other forms of energy very efficiently. The logic operation described in Sankhya philosophy, particularly the five senses or Tanmatras, is a very complex process as if it describes a superneural network that a brain can control through the mind.

Today, I think, machines cannot understand this type of complexity. Machines cannot prepare things with care, respect, or love. These are things that the human brain is capable of doing. Machines can only carry out a specific set of instructions. They can only copy but cannot create their ideas; for example, they cannot create their preferred taste of coffee. The human brain can understand internal forces while machines cannot. The human brain can understand a quantum process without disturbing it. Machines will always measure what disturbs the process, making the no-clone theorem true. Humans use estimation without actually measuring. So I realize that today's classical AI cannot compete with human intelligence. However, AI can be improved through the addition of philosophy based on a nondualistic view that gives a holographic view of the world. We can simplify the operation of an AI-based computer in an artificial vision that can have been initiated based on the dualistic nature of the operation. Vision relies heavily on quantized energy to perform this process. A *chromophore* in a photoreceptor interacts with and absorbs a photon and undergoes a change in structure. This process of structural change is known as 'photoisomerization'. This induces a change in the photoreceptor's structure, and the resulting pathways lead to an optical signal. This work aims to show how a quantum computer can generate holograms, process them, store them, and compare any two random holograms.

Clear Vision: A Very Complex Problem

So what do you expect to see through the so-called Third Eye that is not possible through stereoscopic vision?

1. Clear vision (noise-free), like seeing the flame of fire disregarding the created smoke from the fire
2. An extra depth or dimension using any depth of field or aperture
3. Superimposed images (image reconstruction)
4. Vision beyond duality (duality shows wave–particle at the same time instead of one at a time)
5. Ground state, or singularity, or the end of a vortex
6. Dynamics of a process from resonant tunneling to the protein folding process
7. Ground and excited state simultaneously representing the Purusha and Prakriti
8. All five senses appearing at different frequencies over a wide spectral range as a plant does
    1. **Weak localization and spin-orbit interaction (weak antilocalization) process**
    2. **Holonomic quantum computation**
    3. **MCQI and artificial vision**

In the past, people looked at bright objects in the sky, and many times their vision was obstructed by the clouds in the atmosphere. Reflection often causes poor visibility. This interference effect needs to be destroyed for clear vision. The reflection of the light emitted from bright stars can be produced from the dust of stars, dust clouds, and plasma in deep space, which moves at a very high speeds, and the scattering of the light from the dust causes diffusion, which also happens in a disordered metal for the electron cloud. The electrons can have a very high speed; however, they cannot move in a straight line due to diffusive transport. This diffusive process can result in a coherent backscattering process that lifts the effect of the disorder process partially. This so-called weak localization process is a two-level process, where the forward and the backward coherent waves interfere and produce extra resistance. However, by applying a magnetic field, a phase difference between these two waves can be created that destroys this extra resistance. This is similar to the removal of an obstacle to produce a clear vision by overcoming noise. Here we discuss three techniques: (1) coherent backscattering process, which shows a transition from the weak localization to the weak antilocalization process based on arrangements of superconducting qubits; (2) holonomic quantum computation, i.e., using two qubits to create a four-level system that can be extended by adding a third qubit. A set of three qubits delivering an eight-level system can create a geometric phase. The holonomic gate operation can run these qubits, which can also be implemented in NV centers. This can

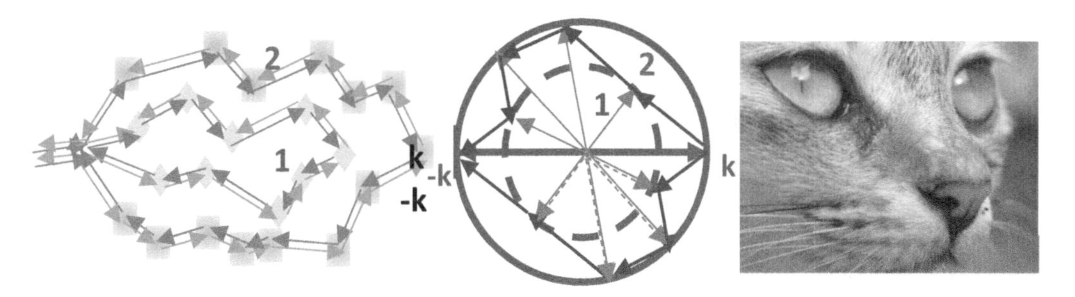

FIGURE 10.3
A double path scattering system with paths going in opposite directions. A diagrammatic representation of coherent backscattering in a disordered system in the presence of a magnetic field. A double path WL model shows multiple backscattering ($-k$ to k) in paths 1 and 2 and a resonance-like interband transition. Clear vision from retroreflection in the eyes.

be used for color images. (3) Multichannel quantum images (MCQI) architecture can be produced from three color qubits.

In Chapter 9, I explained the concept of qubits and how qubits can be arranged for so-called quantum simulation. We have shown how to perform quantum tunneling using qubits. In this chapter, I extend the discussion to coherent backscattering and spin-orbit scattering. Let me start with the weak localization process, which is closely associated with quantum diffusive transport.

Weak Localization Process

According to the modern physical concept, all microscopical quantities possess both wave and corpuscular properties. The mathematical implementation of this idea requires the introduction of the Hilbert space, whose vector describes a state of a system. And, generally, the state can be represented as a superposition of basis states. The idea of superposition finds many implications in different aspects of physics. In particular, it defines the weak localization phenomenon. Indeed, there are very particular paths for a microscopical quantity scattered by an arbitrary random set of scatterers that have the form of a loop. Then, according to the superposition principle, one can construct a superposition of states corresponding to the circumvention of a particular loop forward and backward in time. If states are assigned to electrons, then, since they interfere constructively at some conditions, increasing the sample resistance is expected. The presence of either a magnetic field or spin destroys the completeness of the constructive interference, reducing the effect or even inverting it into a weak antilocalization phenomenon. It should be also specified that the state apparently responsible for weak localization involves two correlated paths of electron and can be simulated through the concept of the Cooperon.

Considering the effects of tunneling, backscattering, and the accumulation of a geometric phase, we see the possibility of simulating weak antilocalization (WAL), in addition to weak localization in a multipath system. We show how a quantum simulator works through the construction of multiple scattering centers in closed paths and tunnel barriers, yielding a large return probability (P_r) for electrons. A combination of inter- and intralayer tunneling in a double-path circuit creates a phase reversal and subsequently the WAL effect.

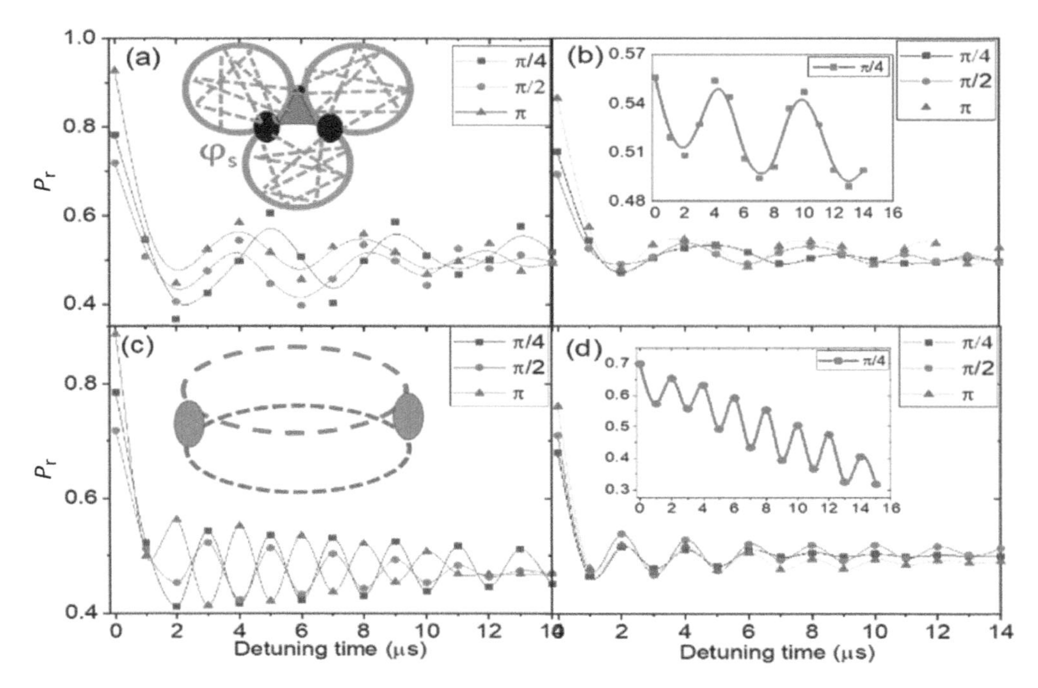

FIGURE 10.4
The probability as a function of detuning time shows an oscillatory feature for inter-layer tunneling. Emulation (a) and simulation (b) of periodic oscillations (Φ_0) and a suppressed WL effect for an A-B ring. (a) Multipath WL model with tunneling applied to a multiquantum dot system forms an A-B phase ϕ_s. (b) Simulation of A-B oscillations without considering WL. Emulation (c) and simulation (d) of $\Phi_0/2$ oscillations for a tube. (c) Multipath tunnel junction forms an A-B tube. (d) Simulating oscillations without WL for a tube. (e) Two-path WL and the A-B ring. The $\pi/4$ pulses and the CNOT gate are used to emulate the geometric phase shift, associated with an A-B.

The connection between the two paths can be established through (1) direct coupling, (2) a tunnel barrier, and (3) a pair of tunnel barriers enclosing space creating an A-B phase. Through this, we also demonstrate the WAL effect and the acquisition of the geometric phase in the circuit.

To illustrate the effect of tunneling on the WL process, we simulate the single- and dual-path tunneling in our circuit, using the single- and two-qubit configurations, respectively. With tunneling induced within the interlayer system, the P_r drops rapidly from its maximum probability at $t = 0$ until it reaches a very low value followed by strong random fluctuation.

Comparing the P_{rr} measured at different angles $\phi = \pi/4$, a difference was found not only at the peak at $t = 0$ but also at the fluctuation point. This doesn't improve intralayer tunneling, which indicates a high level of noise involved in this tunneling process. Extending the work to a closed (triangular) path created by inter- and intrapath tunnel junctions allows for the accumulation of an additional phase, resulting in the WAL effect. The PP_{rr} of the ground state vs. t for both paths shows the WAL effect as the P_r reaches its minimum at $t = 0$.

This is followed by an exponential increase with time until it reaches a stable value of around 0.3 and then fluctuates randomly with a slow increase, which is exactly opposite to the WL effect. We see that smaller ϕ values result in a lower rate of decay. Despite this, our values for ϕ tend toward the same value for P_r, where they fluctuate randomly, which gives the appearance of visible noise in the circuit.

The incorporation of such arrangements of tunnel barriers can add a geometric phase and demonstrate Aharonov–Bohm-type Φ_0 and $\Phi_0/2$ oscillations in a ring and a tube, respectively. In a magnetic field, the phase coherence of two waves is destroyed. The phase change is shown as $\left(\frac{2e}{\hbar}\right)\varphi$, and the factor of 2 arises; $\hbar$ is the waves surrounding the area twice (interpreted as a particle with a 2e charge surrounding the area twice). A whole range of such areas exists. Resistance in such a system oscillates with a flux of = $\hbar/2e$. This type of experiment is performed in a cylindrical film with a magnetic field parallel to the axis. However, for a thin field in a perpendicular magnetic field when the largest phase shift exceeds 1, then cancellation happens due to both constructive and destructive interference. This happens after t = $\hbar/4e$.

Conductance correction in the field H yields coherent backscattering intensity. These features can be explained by the acquisition of the geometric phase.

Geometric Phase

The study of the phase has produced a large number of applications, particularly in holonomic quantum computation by manipulating the circuit parameters where the gates applied on qubits are based on geometric phases. In addition, the acquired phase is made up of the dynamical phase shift, which is related to the energy of the quantum system. However, the pure geometric phase arises from the geometric property of the closed-loop

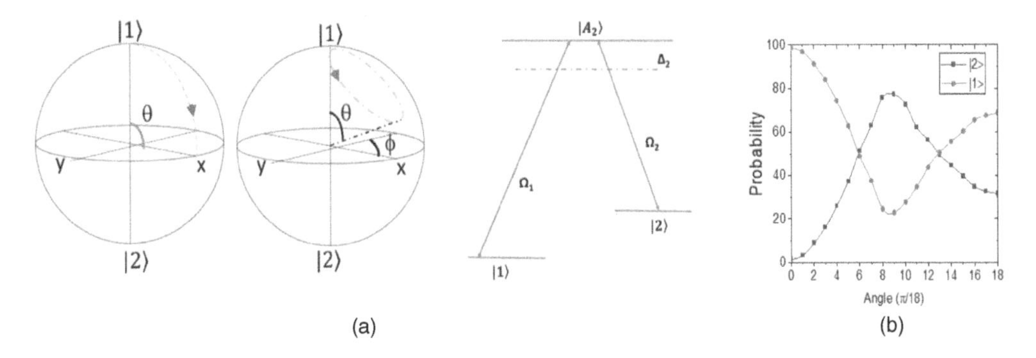

FIGURE 10.5
Simulation of holonomic control on a single qubit. The sphere shows the qubit rotation. Energy diagram for a single qubit showing three horizontal lines which indicate the energy levels. The lines connecting the levels indicate the transitions. The probability shows an oscillatory feature. (a) The rotation path is implemented by a single holonomic gate. In the first diagram, the qubit is rotated about the y-axis by $\pi/2$. Since the path forms a closed loop, a geometric phase can be created via a single iteration. The NV center consists of a triplet ground state. The detuning is between the ground and excited state since these two states are the states that were used for the simulations. The detuning is between the exited state $|A_2>$ and the ground state $|1>$ and $|2>$. Δ represents one photon detuning frequency. (b) Gates with the angle θ variable after initializing the qubit to $|1\rangle$. The probabilities of the final states are to be measured on the $|Z\rangle$ basis from both $|1\rangle$ and $|2\rangle$ and plotted against θ. The red and black lines indicate the behavior of the qubit in the presence of noise.

or path that the system follows in the Hilbert space. The geometric phase, dependent on the area enclosed by the loop, can be extremely resilient to the effects of noise since noise does not cause any changes to the area enclosed by the loop. In theoretical physics, phases are needed to understand the Aharonov–Bohm (A-B) effect, which can be used to show that the wavefunction of a particle is affected in the absence of an explicit magnetic field by scalar and vector potentials. The A-B effect happens because the wavefunction of a charged particle is coupled to an electromagnetic potential that can be described as an instance of a geometric phase. The geometric phase, or Berry phase, is acquired by the adiabatic evolution of a quantum system around a circuit (path of rotation). This can be generalized to nonadiabatic and nondegenerate cases. If the energy levels of a system are degenerate, the cyclic evolutions of the degenerate subspaces will produce a path-dependent transformation that is called the 'non-Abelian phase'. Conversely, if the energy levels of a system are nondegenerate, the phase will be Abelian. In a qubit, this geometric phase will have a nonzero solid angle associated with it. The solid angle can be used to store information about the acquired geometric phase. The trajectory of the unit vector along the Bloch sphere is equal to half of the solid angle. Since the Berry phase is only dependent on the global geometry of the loop and is resistant to small errors, quantum gates that rely on the geometric phase have higher performance than other types of gates. Such quantum gates can be protected against the effects of decoherence that may improve the performance of such gates. We shall see how the geometric phase is connected to spin-orbit coupling (SOC), which can be simulated by a quantum computer,

In Chapter 9, we presented space–time distortion as a very general idea that works in a black hole or a quantum vortex. The generalized idea of this effect is space–time distortions associated with the curvature of the space compared to flat space. For example, the introduction of a pair of pentagon-heptagon breaks the hexagonal symmetry of flat graphene sheets by adding a curvature vertical, which is a lateral space–time distortion. Further, distortions happen by vertical space–time distortion in a vortex structure, which adds helicity to the space. In both cases, an excited (p-) state is created from the distortion of the ground (s-) state. The height of the distortion or the helicity corresponds to the level of excitation of the p-state. More breaking of symmetry creates more p-like states, hence producing stronger spin-orbit interactions in a disturbed system.

Color and Geometric Phase

Color is produced from inelastic scattering that can be seen from Raman spectroscopy. The color of different samples corresponds to the frequency shift of incident light, which depends on the polarization of the system (atoms or molecules). An electric field creates the polarization and introduces spin-orbit coupling to the system. A geometric phase shift gives color dependency of the excitation level measured from the Raman spectroscopy. The geometric phase is a means of excitation of the energy level of the excited state. A larger loop means stronger SOC. So the frequency shift or the color can be registered by the geometric phase. The geometric phase is not spoiled by random scattering. It protects the coherent backscattering so that the phase memory is retained. The rotation of the qubits can be affected by random scattering temperature fluctuations, magnetic fields, or other decoherence processes. However, the geometric phase protects the rotation of the qubits in a topological manner. Quantum computation reaches the highest accuracy. By introducing

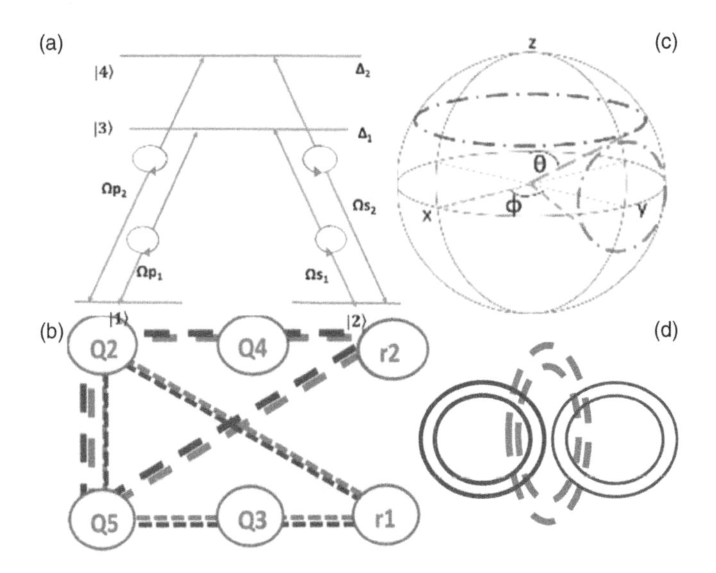

FIGURE 10.6
(a) A four-level energy diagram of the spin triplets of an NV center having a complete nine levels showing the Rabi frequencies; the detuning is $\Delta_1 = \Delta_2 = \Delta$. (b) The holonomic control on a single-qubit involves a rotation path forming a closed-loop (a geometric phase). (c) Qubit arrangements show a cross-linking of qubits assigned to pseudospin and orbit. (d) Two spins and orbits in a double-path system. The sphere shows the holonomic rotations of the qubits which creates an area or a geometric phase. The intersecting loops represent spin-orbit couplings.

a defect center or a magnetic impurity, the spin-orbit coupling can be explained, the coupling strength records the geometric phase shift, and the color of the image that needs processing. The images can be recreated accurately through the knowledge of the geometric phase very accurately. The geometric phase can be recorded through holonomic operations on the qubits, which are elaborated on in the next section.

Emulating Spin-Orbit Interactions in Multipath Weakly Localized Quantum Systems

Emulating magnetoresistance (MR) associated with spin impurity scattering has been attempted by superconducting qubits in triangular lattice-like configurations. Originally proposed as a relativistic effect spin-orbit interaction (SOI) has been studied as one of the most important problems in condensed matter physics, topological materials, and condensates. Strong SOI can overcome the effect of disorder in coherent backscattering, so-called weak localization (WL), which becomes a weak antilocalization (WAL) effect with a minimum of the WL peak as observed in field-dependent magnetoresistance (MR).

In previous chapters, we have discussed the influence of the SOC on superconductivity exhibiting a spin-triplet state. The SOC can be explained by a pair or a dual state that works oppositely to the Cooper pairs. In SOC, the rotation of spins creates a geometric phase in addition to any other phases. The creation of the geometric phase can affect the scattering process like the application of a magnetic field that can explain the WAL process. Let us explain the weakly localized process as the antiparallel arrangement of a pair of spins. The strong interaction between these two spins produces a small energy gap so that the current-carrying particles can overcome strong localization (or a deep potential well).

Since the time-reversal symmetry is not broken in this scattering process, no geometric phase can be generated. The process may be depicted as two rings (zeros) rotating in opposite directions. To add a geometric phase, we should introduce another ring or an orbit normal to the plane of the other two rings. Such a configuration produces a unique geometry, i.e., a solid angle based on the angle and the area traversed. The split of one circle into two (orbits) defines duality, whereas the third cycle (representing a spin) establishes a 'trinity' that not only split two cycles but also keeps the third cycle independent of the two. This adds an extra dimension or an additional space that can be seen as a central peak in the energy gap of a superconductor transforming it into a gapless superconductor. The competition between the spin and orbit is described as a state of different superposition or orientations of the cycles. It takes all possibilities, i.e., two cycles rotating in the same direction (both clockwise and counterclockwise) and also in opposite directions, which can also be combined. As a result of the superposition, we see satellite peaks on either side of the central peak: in total, three peaks. This structure reminds me of the model of god as the center supported by two associates. In this process, the central peak is protected by the side peaks from any disturbances. In the next subsection, we see that the central peak can be split in the spectra due to a holonomic operation that resembles the SOC. The geometric phase works like Maya, which protects the center or Purusha from disturbances. A geometric phase can be produced from a set of three qubits or three spinors.

Spin-triplet superconductivity in the presence of RSO at the interface of heavily boron-doped diamond has been claimed to show satellite peaks in the MR spectra. However, the hallmark features of SOI, i.e., the satellite or split peaks of MR (in addition to the central peak for WL) have not been properly explained theoretically or by simulation, which is now presented.

Earlier in the interface, the SOI effect was described by a double potential well system like a vortex or an Ahanonov–Bohm (A-B) ring topologically protected from the scattering effect. In the Rashba model, the strength of the SOI depends on the Coulomb potential, which binds two opposite spins through a repulsive interaction that can be mapped with two qubits incorporated in a multipath WL system created by other qubits. In this subsection, this effect can be created by introducing an AB phase created through multiple tunnel barriers using multiple qubits in a double-path WL system. Here we control the rotation of qubits precisely, such as fixing one qubit while rotating two others, forming a triangle in a Rashba-type configuration. Features of spin-triplet superconductivity produced from the SOI, particularly the split of the interference peaks of the AMR features, can be experimentally demonstrated by a quantum emulator. In the Rashba SOI model, the strength of the SOI depends on the Coulomb potential, which binds two opposite spins through a repulsive interaction. This can be described by an excited-double four-level system (|1>, |2>, |3>, and |4>), which we associate with four qubits (or bilevel systems). This can be mapped with two qubits incorporated in a multipath WL system created by other qubits, i.e., by transforming a spin into two pseudospins. The states |1> and |2> correspond to pseudospins $\pm 1/2$, and their interactions are spin-like. A spin-qubit has been defined as the ground state Kramer's doublet. The interactions of states |3> and |4> are bosonic-spin-like or orbit-like. Most importantly, interactions between levels |1> and |3> or |4> {or between |2> and |3> or |4>} correspond to spin-orbit interactions. This effect can be created by introducing an Aharonov–Bohm (A-B) phase created through multiple tunnel barriers using multiple qubits in a double-path WL system. The pump and the stokes field with frequencies γ_{p1}, (γ_{p2}), and $\gamma_{s1}(\gamma_{s2})$ correspond to pulses applied to the qubits with the Rabi frequency Ω. These two pairs of paths correspond to the coherent backscattering or the weak localization orbits.

Here we control the rotation of qubits holonomically by creating a closed path. This can be achieved precisely by fixing one qubit while rotating two others, forming a triangle on the Bloch spheres in a Rashba-type configuration.

The SOI effect on WL showing a local minimum or split peak in P_r is presented. By adjusting the strength of the SOI, the P_r peak associated with the WL can be split, yielding a new peak or going to a minimum similar to the WAL effect. In the SOI, a spin revolves around a fixed spin or orbit, creating the space in between, which is similar to creating an AB ring. Therefore, we construct a system for a two path WL (T-C model) without any tunneling between the paths, which protects the inner part from dephasing in a topological manner. They are separated by an AB-like ring created from a square arrangement of spin centers or qubits, which is stabilized by pulses fired from a detuning box. Considering the WL paths creating an AB-like ring, we describe the RSO effect as the split of the ring, which can be verified from the rotation of the magnetic field or the qubits.

Holonomic Quantum Computation through Dark States

The study of phase has produced a large number of applications, particularly in holonomic quantum computation, by manipulating the circuit parameters where the gates applied on qubits are based on geometric phases. The geometric (or Berry) phase is acquired by the adiabatic evolution of a quantum system around a circuit (path of rotation), which can be Abelian if the energy levels of a system are nondegenerate. Conversely, if a system's energy levels are degenerate, the cyclic evolutions of the degenerate subspaces will produce a path-dependent transformation called the 'non-Abelian phase'.

In a qubit, this geometric phase will have a nonzero solid angle associated with it. The solid angle can be used to store information about the acquired geometric phase. The trajectory of the unit vector along the Bloch sphere is equal to half of the solid angle. The pure geometric phase arises from the geometric property of the closed loop or path that can be extremely resilient to the effects of noise since noise does not cause any changes to the area enclosed by the loop. Since the Berry phase is dependent only on the global geometry of the loop and is resistant to small errors, quantum gates that rely on the geometric phase have higher performance than other types of gates. Such quantum gates can be protected against the effects of decoherence, which may improve the performance of such gates.

Holonomic computation can be performed on qubits implemented in NV centers or superconductors. However, the NV center is seen as a more promising candidate for the implementation of holonomic computation because (1) its quantum state can be manipulated by laser pulses or microwave fields, (2) it has a high fidelity at room temperature, and (3) it has a long-lived spin-triplet. The lower levels of NV centers can function as qubits. Dark states can also control the level of decoherence: The phase difference between the dark state of a holonomic gate and the initial state of the qubit determines the decoherence of the system.

A dark state is a state of a three-level atom that cannot emit or absorb photons. Dark states have an eigenvalue of zero, and dark states of Λ-three-level systems have been used for quantum memory storage. Dark states are a special case of dressed superpositioned states as such states do not interact. It is like a bound state. Dark states are needed in gate operation to remove dynamic phase shifts. This allows the proper measurement of the geometric phase.

We have been experimenting with the IBM quantum experience technology to determine the capabilities of simulating holonomic control of NV-centers for three qubits describing an eight-level system that produces a non-Abelian geometric phase. The electron return probability can exhibit spin-orbit-coupling-like behavior, as observed in topological materials based on the extra geometric phase. A three-qubit system produces a similar effect to Rashba spin-orbit coupling (RSOC) and vortex structures in materials that are highlighted in the chapter. A holonomic gate can be defined as an operation or gate that causes a qubit rotation, such that the qubit acquires an Abelian phase that can be implemented on qubits in optical systems such as NV centers or superconductors. For the three-qubit system in a trigonal arrangement that occupies a closed space, we can generate a non-Abelian geometric phase and show the superiority of off-resonant holonomic gates over the on-resonant gates. This development shows particularly interesting potential as NV centers are viable qubits at room temperature, and the operation of universal gates on such qubits is a step forward in the development of universal quantum computers. In the second scenario, we consider a similar configuration for the system; however, we measure the effective rotation time for the varying values of Δ_1, Δ_2, and Δ_3. We measure the rotation time of each of these paths, as well as the effective rotation path to obtain the P_r of the state $|1\rangle$. Each rotation accumulates a geometric phase that is Abelian, so that $d\varphi = 0$ for the first two rotation paths, and only either $d\varphi$ or $d\theta$ is nonzero for the whole rotation. Although each separate rotation represents Abelian geometric phases, the effective phase can, however, be realized as non-Abelian. In qubits, a dark state is a state where a qubit undergoes trivial dynamics (i.e., the qubit can be described by the qubit Hamiltonian). For the creation of geometric phases, the rotational path traversed by the qubit is usually a closed loop and can be implemented on an arbitrary axis. Dark states can be created by the strong overlapping of two orthogonal states like the spin-orbit interaction. In condensed matter physics, it is well known that the geometric phase arises from the RSOC. This, however, has not been experimentally demonstrated using a quantum simulator that requires at least three qubits. Earlier, a three-qubit system was employed to generate a synthetic magnetic field, a chiral ground state current, or a chiral spin state; however, the geometric phase was not captured, which was attempted in the present work. The purpose of the holonomic operation is to create a geometric phase through a closed loop of multiple iterations. Dark states can be described as the bound states or the vortex core, which arises from the strong spin-orbit coupling, which results in weak (anti-) localization phenomena without breaking the time-reversal symmetry. This picture becomes very interesting for three vortices that can be controlled by a single operation. We see fine control of the P_r with the Rabi frequency (or RSOC strength). At zero frequency, P_r has a peak at the origin or zero detune time (similar to the weak localization phenomena observed in the magnetic field-dependent resistance of metal with the effect of RSOC). This becomes a minimum by the application of the Rabi frequency, which is similar to weak antilocalization. This is a hallmark feature of RSOC in a topologically protected material. While two holonomic operations perform a weak localization-like feature, the third holonomic gate breaks the time-reversal symmetry. Hence the operation of a three-qubit system works more efficiently than a two-qubit system.

The one-qubit system can only be encoded with the ground states of the NV centers (or with one dark and one bright state). This means that only one dark state can be measured. However, since the ground states are degenerate, no transitions are possible. For the three-qubit system, since there are more dark states, transitions between the dark states can be observed just through the holonomic control of one of the three qubits in the system. The transition is also quite interesting because it causes the amplitudes of the states other than the initial states to increase for a short duration. Also, with three-qubit systems, the system can exhibit geometric dependence, which is not possible for one- and two-qubit systems since only one configuration exists for these systems, while there are two configurations for the qubits in the three-qubit system.

Optimization and Image Processing

Optical techniques to excite a large number of NV centers and the readout will be by electrical means as a resonator (superconducting qubits). Holonomic operation in flux qubits has limited scope, unlike operation in spin qubits (three NV centers or qubits will be employed for holonomic control of NV centers for a three-qubit system, plus a combination of SC phase slip qubit [flux] and NV centers). Add intra- and interlayer tunneling for image processing. Image processing and optimization need a feedback process (reinforcement). Three qubits are needed for (RGB) color coding. The feedback process is possible only for spin-triplet states, which are caused by a bound state (Andreev or Kondo) or dark states, which can be created in NV centers (real or simulated). These dark states may be connected to form a neural network.

MCQI for Artificial Vision

Vision is the ability to create a hologram from a superposition of two input images and then to process this hologram without any other new inputs. This is similar to the reconstruction of images by the brain after closing the eyes. What is needed for this hologram? Two or more multichannel quantum images (MCQI) (one is the original image, and the others are the transformation of this image) are needed, which is superior than the flexible representation of quantum image (FRQI) processing. We can then superpose these images creating the hologram. This will create a three-dimensional image that has not been processed by quantum computers before.

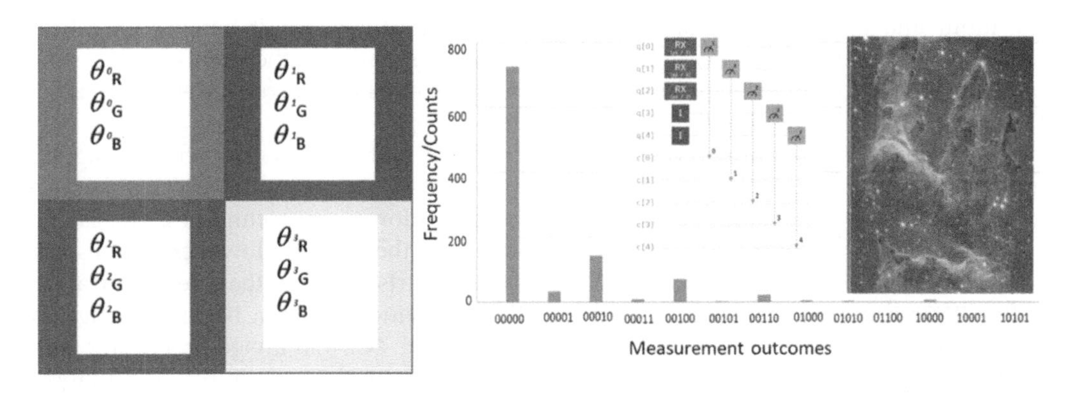

FIGURE 10.7
(Left) The multi-channel representation was proposed to capture the RGB channel information. This is accomplished by assigning **Three qubits** to encode color **(middle)**. A square is divided into four smaller squares. Each square indicates the color and the position of a point in the image. **(Right)** Pillars of creation as observed by the James Webb space telescope. Quantum image processing will collect more information from the photos. Eagle nebula shows pillars or finger-like structures made from clouds and a large number of stars.

Like cooking or preparing food from a mixture of materials, we mix images to create the best-looking one. First, we will need to generate the hologram just mentioned. We will consider using two or three images. However, a quantum computer cannot directly process 3D images. To solve this problem, we will have to blend these images. This can be done by using a blending operator that can blend foreground and background images. The blended image will still be defined by the MCQI equation and also with the same color and position as the original hologram. We will create lots of holograms or false vacuums using this method. Then we will encode the holograms (now 2D images) into a strip. This completes the preparation stage of the experiment. The next stage is to transform the color information on the holographic, images that can be achieved through the simultaneous operation and swapping of quantum gates. This will then be fed into the superconducting circuit as the environment states with three qubits that can control three basic colors: red, green, and blue.

The first stage is to create an artificial eye or a camera that can focus on an object with intelligence to take a lovely picture. This focus creates a point that can then be extended into a line, which is further extended to become a flat plane. This plane is the image that can store information. However, a 2D image does not store enough information (it lacks a *z*-axis). This is when we create a hologram by creating layers of images. This can be done in other ways as well. One is by translating the image in a perpendicular direction, which leaves a 3D space with a tapered end like a funnel or a vortex. Another way is to view the image through a red and blue filter simultaneously, which creates a superposition of these views. A quantum computer cannot directly process such images. We must therefore create a 2D image. This can be done by using a quantum Fourier transformation (QFT). This collapses all the images into the desired image size. Now this can be processed using standard image processing techniques. When you fold a 2D object, it creates additional space that can be used to store information like a fist. Now you use two hands (two potential wells). You want to transfer energy between these two wells. This has been proposed by the photosynthetic protein folding model. In this model, photosynthesis is described as the switching of proteins. The flaw: no superposition of states. The superposition of states creates dreams. When you close your eyes (blinking or peeping), the information circulates in your brain like a feedback loop. In a quantum circuit, this can also be achieved as only one input image is required for processing without any new input images. Quantum circuits, however, lose this information due to measurements and the resetting of the circuit. The circulation creates a unique superposition like a supercurrent. Hence I create a Third Eye for intelligent vision.

AI and vision can be improved through interactions with the environment, i.e., agent, environment, and register interaction. It is nothing but SWAP gates and CNOTs that are run through a reward protocol, as used in quantum reinforcement techniques. This is like a quantum feedback loop that increases the fidelity of the circuit. However, it does not use disorder or random scattering, which are integral parts of vision that are significantly different from normal or elastic scattering. By employing quantum backscattering, an image that is diffused and not so bright can be analyzed. Normal image processing uses FRQI and MCQI techniques. We employ two sets of three qubits (six in total) to analyze complex images. The input is used as the agent (S = In), as the channel of interest operator (CoI). Then the environment is used as the strip and the register as a set of three qubits. Now for the three qubits, we used ISWAP or simultaneous ISWAP gates instead of CNOT gates. This ISWAP gate can be represented as a chiral SWAP or ASI gate. So the CNOT is represented by Toffoli and Fredkin gates, which will improve the fidelity of the circuit. This will solve the problem of weak localization and quantum backscattering. Due to the scattering process, the color of an object appears to change when it is received by the human eye. Our eyes and brain can be confused while processing complex images. It fails to distinguish the desired object if it is mixed with a group of objects, which poses a major problem in astronomical observations and medical science. AI vision is even worse since it cannot distinguish between living and dead objects.

How do you form an image? It is through interaction with nature, an exchange interaction like a SWAP gate. But we need a simultaneous ISWAP gate that will exchange interactions among three completely different elements. The third one covers the first two as Tamas. This model is like supercoiling. Red and green will interact and swap positions, and then blue comes as the outer layer, which swaps with the combination of RG. Consider light, darkness, and halo. Light and halo interact by swapping their position. This is swapped with darkness. We see this happening in nature with the appearance of the Sun in the form of dawn to Moon and dusk. The entire process is covered by the night or darkness, which also interacts with the Sun. Any element in addition to these three can be considered a further swap process. Halo is described as Maya that interacts with light or darkness. Halo is produced from random backscattering, a diffuse process of producing an image from light. The probability distribution of simultaneous swapping can describe the image, which can be compared to another probability distribution or image. In other words, the tomography of two images can be compared. The generation of tomography is vision. This can be developed with three different qubits, a combination of superconducting flux qubits, and NV centers that can produce a full RGB color. We create a 3D profile on a flat space. We can create a nice 3D profile like our face and then compare it to another 3D profile.

Synergetic Quantum Computing and Image Processing (Vision)

Quantum image processing, which deals with representation, processing, storing, and comparing images, has been seen as an important part of quantum computing. Representation is a fundamental aspect in determining the processing technique that will be used. There have been many proposed methods of representing quantum images, including Real Ket, qubit lattice, FRQI, and MCQI. FRQI images are the most extensively studied out of these representation methods as it consists of only one color qubit. Multichannel quantum images (MCQI) were introduced to represent full-colored images, but since the only difference between MCQI and FRQI images is color, the MCQI image is an extension of FRQI images. However, these images take up a lot of qubits, which is why the trend moved away from them. Applications of QIP include stenography and quantum films, as well as their processing and readout. Most of these applications are based on FRQI images.

Quantum computers can play a big role in helping AI to think better and faster and hence be smarter than the human brain. But just as human brains have to grapple with random data, AI and quantum computing both face the issue of information entropy. We can see how a philosophical concept can be geared toward understanding the nature of entropy in both AI and quantum information, together with experimental research toward the reinforcement of entanglement and quantum coherence as a complement to error correction.

Currently, those working in quantum computing appreciate the complexity and difficulties of the development of quantum computers and quantum communications. It is our opinion that human agency alone is not capable of developing high-efficacy quantum information and that an AI agent can codevelop quantum computers.

We try to capture the image of nature, an effort that contains uncertainties. A piece of cloud covering the Sun (or a distant star in a gaseous nebula) destroys information about the source due to random backscattering of light by the particles in the cloud. A coherent backscattering process creates interference that can give false information about the source. A strong electric field can destroy the interference by breaking the cloud and allowing light to propagate over a large distance and carry the right information from the source. This is known as weak antilocalization, which can result from strong coupling between the electric field and the orientation of the aligned particles (spins) in the cloud. The rotation of the spins associated with the particles may be controlled by the polarization of light, and the wavelength of a spectrum of light is registered through the rotation. Now all information must be protected by squeezing the states. This can be done by maximally entangling the states. This is equivalent to focusing the light at a point that can form a sharp image. This focal point works as the core of the vortex. In a quantum computer, such a state can be obtained by a few qubits or atoms through maximum entanglement, which is equivalent to a 'black cat' that contains all information of the source. This state is neutral and static, a dormant state that does not emit any information (as a grand state). From the maximally entangled state, heat is produced, which also creates polarization due to instability, followed by an electric field. This produces an excited state (space) that emits radiation. Hence the focal point is shifted and various frequencies (colors) are produced. The core of the vortex splits into two halves. In this way, we explain creation.

Summary

A large number of researchers used quantum simulators and found as a very useful tool for simulating quantum many body systems which improves the understanding of coherent backscattering and accumulation of the geometric phase. Recently, the potential for simulating holonomic control of vacancy-related qubits has been used for image processing. Using holonomic operations, the color of the images can be restored properly. The philosophical view of vision is connected to machines and quantum technology.

Conclusion

Science is a special type of knowledge which deals with the change on expansion of an object. It constantly searches for truth. Scientific tools are used to find the smallest entities, very far distant objects, and very complicated structures. The smallest objects can be elementary particles, atoms, nanostructures, and biological materials that carry life. Various microscopes with high magnifications were developed by using optical lenses, electron lenses, and even electron tunneling (scanning electron microscope), which can identify atomic or subatomic structures. With the invention of the atomic force microscope, earlier predictions of hexagonal microstructures of graphene can be matched. However, until today, some important subtle objects related to our life such as DNA structure cannot be seen directly.

One of the most important parts of science is to search for life in the universe and to find a model of creation through the invention of an astronomical device that can magnify the images. Telescopes provide a means of collecting and analyzing radiation from celestial objects even those in the far reaches of the universe. Thus telescopes have opened our eyes to the universe, revealing geography and weather on the planets in our solar system. An important factor for magnification is the focal length of the telescope, which was 1,100 mm for Galileo's telescope and which could magnify 30 times. A longer focal length will shorten the field of view but increase magnification, which is ideal for observing planets and the Moon. Galileo's telescopic observations illustrate the advanced mode of research about how a tool can see and collect evidence by changing the understanding of the cosmos. Galileo thus turned telescopes toward the heavens.

The ingredients of our life are spread throughout the universe. While Earth is the only known place in the universe with life, detecting life beyond Earth is a major goal of modern astronomy and planetary science. Later, for the study of exoplanets and astrobiology, the James Webb Space Telescope was developed by NASA to measure the chemical makeup of atmospheres of planets around other stars. The hope is that one or more of these planets will have a chemical signature of life. To detect life on a distant planet, astrobiologists will study starlight that has interacted with the planet's surface or atmosphere. If the atmosphere or surface was transformed by life, the light may carry a clue called a 'biosignature'. James Webb telescope has a focal length of 131.4 meters and the capability of observing objects that are 13.6 billion light-years away from us for an explicit detection of life. However, it is challenging to decide which planets may host life, since the unbounded oxygen, the strongest signal for life, may not be determined by the telescope. Telescopes of the future could be designed to find life on other planets. Scientists are studying waves to block bright light coming from a planet's host star in order to reveal starlight reflected from the planet. Three such ground-based telescopes are currently under construction: the

DOI: 10.1201/9781003304814-12

Giant Magellan Telescope, the 30-meter telescope, and the European Extremely Large Telescope.

The big question is, 'Is there actually life beyond Earth?' We do not have a universally accepted definition of life itself. That said, we might not need one. We need to detect only the telltale signs of life in an exoplanet atmosphere, and we have a better understanding of what those are like here on Earth. Life must turn up in our neighborhoods: beneath the Martian surface, perhaps, or in the dark subsurface oceans of Jupiter's moon Europa. Maybe the dream of the ages will come true, and we can eavesdrop on the communications of extraterrestrial civilizations. Evidence of technosignatures or traces of technology might be captured. Barring these strokes of luck, the job may be harder. Light will be the key—light from the atmospheres of exoplanets split up into a rainbow spectrum that can be read like a bar code. This method, called 'transit spectroscopy', would provide a menu of gases and chemicals in the skies of these worlds, including those linked to life. They dwell in the caustic chemical pools of Yellowstone National Park, in the dry valleys of Antarctica, in the superheated vents of the ocean floor, and they belong to branches of life that split from ours billions of years ago. Extremophiles, which are forms of life that love extreme environments, thriving in conditions that would kill anything else, might also be analogues for strange life on distant worlds.

While small and distant objects can be visualized through the increased magnification of the telescopes, the images are found to be very complex since they are covered by gases and faint objects lose their intensity. Recently, photos of black holes in our galaxy were published, which is achieved through improved image processing techniques. We need better image processing techniques to understand signals and to employ quantum artificial intelligence to remove noise from the data. We look for more extraordinary photographs of the universe, which can develop a model of the universe and creation (God!).

Science vs. Belief of the Primordial Energy: Shakti

There is a common belief that science and religion do not support each other. They do not have the same targets, and they may be contradictory since religion is based on belief in an invisible God and science is based on observation of real objects. Eastern culture has tried make both sides converge through careful observations. God is intellect within our own body. The Indian sages described the physical laws of nature through practical examples like oceans, snakes, pots, threads, etc. We do not know who discovered the meditation process to control the mind but have always looked at a strong attractive force that controls the universe. This force field is universal and described as God. Force cannot be seen, but its effect can be manifested. Some great religious practitioners described the force or energy from direct observation as if it maintains the equilibrium in the universe. This energy creates space and time. In more recent times, some great philosophers tried to make direct comparisons with the laws of physics.

Despite the approach of unifying science with religion, many problems of civilization have not been resolved. A religious scholar cannot or does not like to understand the experimental laws of nature. Similarly, a scientist disregards anything that is not explained by law or formula. In both cases, a factor seems to be missing. It was assumed that nothing is certain or immortal (except one constant); however, to attain a state of certainty, there is always an attempt through burning out all uncertainties by igniting the fire of knowledge.

Understanding that fire is protected in a closed space, its liberation can be imagined in an egg and breaking it to see the red part inside. It is as if space is holding the red-hot Sun, which delivers its maximum power to others. This is demonstrated by sages and Yogis most notably in the form of Half opened eyes. This is a state of vision that is neither closed nor open (half-closed) and that acts as a slice of light or fire, which is particularly seen through the orifice as in red-hot burning wood. This can be opened as the lightning in the sky in a dense cloud. Cloud is meditating or gaining energy through sleep.

Ancient philosophy was developed thousands of years ago before Newton or Kepler. Indian astronomy has a long history, stretching from prehistoric to modern times. Some of the earliest roots of Indian astronomy can be dated to the period of the Indus Valley Civilization or earlier. Astronomy later develops as a discipline of Vedanga or one of the auxiliary disciplines associated with the study of *Vedas*, dating to 1500 BCE or older. The oldest known text is the *Vedanga Jyotisha*, dated 1400–1200 BCE (with the extant form possibly from 700–600 BCE). Astronomy in those times was essentially the science of time determination without the help of any precise instrument. It was centered around the Sun and the Moon as well intended to study the natural divisions of time caused by the motion of the Sun and Moon, such as the days, months, seasons, and years. Special attention was paid to the study of the times of occurrences of new moons, full moons, equinoxes, and solstices. There are many references to the Sun, Moon, stars, planets, meteors, etc. in Vedic literature before 1500 BCE. The Sun is the Lord of the universe, and the Moon shines by the light of the Sun. The Earth is described as a sphere. Even casual observations of the sky would reveal that there are three clear time markers in the sky, namely a day, a lunar month, and a year. All the major civilizations tried to understand the correlations among these units of time.

The modern atomic shell model is merely an improved version of the ancient atomic or Paramanu model developed several hundred BCE. Mass–energy equivalence was described as the basic model of God. It is surprising to notice that no unified model, either in philosophy or in science, has been developed, although a holographic universe is modeled. Although India was great in mathematics and the ancient scriptures were written based on some astronomical observations and measurements, no mathematical treatment has been given to support the rules or laws of nature, or at least they seem to be hidden. In this book, we first describe the philosophical models and compare them to the most famous equations in physics such as Newton's laws of motion, Maxwell's equations, and Einstein's relativity theory. We unify these equations into a vortex model and the interaction of vortices.

A vortex model of the universe that can connect the classic and the quantum worlds is presented in the book. A vortex can range from a microscopic scale starting from a point to a macroscopic scale that is divergent and expands on the astronomical scale. Two vortices can merge and interconnect through tunneling, which forms a wormhole. A vortex can also be split into different parts. A vortex creates space surrounded by a rotational field that is not accessible by any external fields. A vortex can reject external (magnetic) fields like a superconductor. A vortex has a layered structure as if circular planes are separated by intermediate space, which creates discrete or quantized energy levels. The interior of a vortex is maximally chaotic or disordered and can store the maximum amount of information. A vortex can be described as a combination of electric and magnetic fields that develops a spin-orbit-like coupling. The core of a vortex is an orbit (charge-like), which is surrounded by spin liquid. It separates the quantum fluid or spins liquid by pinning the charge states. The core of a vortex forms a bound state or a dark state with zero eigenfunctions. A dark state can store a very large amount of information and can be used as an ideal quantum memory. Since a vortex is created by spacetime distortion, it is a playground for studying the disorder. Thus a holographic (multiverse) can be created not only by theoretical means but also by experimental simulations. We employ a set of quantum bits (artificial atoms or zeros) for the simulation. In this way, we can connect 'zeros' through addition, multiplication, and even division.

We conclude this book by discussing vision.

Vision by Overcoming Dualism

Having created the world, the creator would like to look at his creation, as a painted and colorful earth. Like space–time duality, the onlooker (creator) is the subject (I) looking at the painting (creation) the (object), which is embedded in this experience or what this experience is all about. Almost immediately, the painting also evokes emotions in the subject, thoughts, ideas, and understanding. The feelings generated are qualitative and subjective to the onlooker. The onlooker is called the 'Pramata' or 'Knower', and the painting is the 'Known'. The relationship holds the knower and what you know. These are the possible levels of direct or indirect cognition, but the subject–object binary is the primary feature of the nature of experience with interactivity between the Knower (experiencer), the Known (experienced), and Knowledge (the instruments of knowledge). *Advaita Vedanta* clearly defines consciousness by saying that anything you call 'this' in your experience is an object (insentient) and that which experiences 'this' is called '**consciousness**'. This demarcates what consciousness is and what it is not. The eyes are the seer (the first contact point of this experience), and the form of the object is the seen or experienced. The seen or objects seen can vary or be relative to the seer, who is composite or whole (still relatively speaking). The painting can be replaced by something else, but your eyes remain relatively composite. And one is conscious of the eyes, one knows its conditions, whether they are closed or open; at a higher level, the eyes are known, and the mind is the seer. The entire painting is experienced in the mind, or cognized by the mind, with the aid of the instruments of knowledge (the senses). Since the contents of the mind (thoughts, ideas, feelings, memory) are known, then *Advaita Vedanta* says that at the highest level the mind is known by the witness, or Sakshi, and is not known by anything else. This introduces us to the Sakshi, or Pure Witness Consciousness, the real protagonist of the play, the real knower who is not known by anything. Form is perceived, and the mind is the perceiver, the mind with its modifications is perceived, and the witness (the Self) is verily the perceiver, but it (the Witness) is not perceived by any other. This Self is the core of our self and is also the ground on which experiences take place. The distinction between the Sakshi/Witness Consciousness and the Knower/Pramata is the key to understanding the reality we experience in terms of consciousness. In *Advaita Vedanta*, the reality of the world is the nondual Self, whereas the Other is just a projection or appearance of this in the ground of one consciousness called 'Brahman' or 'Atman'. Mind (subtle matter) and Matter are therefore projections of this nondual Self in itself. There is no second reality apart from Brahman or Existence-Consciousness-Bliss (Sat-Chit-Ananda). It is the irreducible pure subject that is the only reality and independent of anything. It appears as two: the Self-Other binary but also the One Existence-Consciousness-Bliss. The Space–Time Distortion vanishes at this point.

ॐ अपवित्रः पवित्रो वा सर्वावस्थां गतोऽपि वा । यः स्मरेत्पुण्डरीकाक्षं स बाह्याभ्यन्तरः शुचिः ॥

Srimad Bhagavatam 6.8.4–6

"OM Apa-with Pavitro Vaa Sarva-Avasthaam Gato-{A}pi Vaa Yah Smaret-Punnddarii-Kassksam Sa Baahya-Abhyan-Tarah Shucih"

Om, if one is Impure or Pure, or even in all other conditions, as soon as they remember Sri Vishnu with Lotus-like Eyes they become Pure outwardly as well as inwardly.

Definitions of Scientific Terms

VECTOR: Vector is a quantity that has both magnitude and direction. It is typically represented by an arrow whose direction is the same as that of the quantity and whose length is proportional to the quantity's magnitude. Although a vector has magnitude and direction, it has no particular position.

MASS: Mass is the quantitative measure of inertia, a fundamental property of all matter. It is, in effect, the resistance that a body of matter offers to a change in its speed or position upon the application of a force; the greater the mass of the body t, the smaller e change produced by an applied force.

TIME: Time is the continued sequence of existence and events that occur in apparently irreversible succession from the past, through the present, and is a component quantity of various measurements used to sequence events, compare the duration of events or the intervals between them, and quantify rates of change of quantities in material reality or the conscious experience—time is often referred to as a fourth dimension, along with three spatial dimensions.

VELOCITY: The velocity of an object is the rate of change of its position concerning a frame of reference and is a function of time. Velocity is equivalent to the specification of an object's speed and direction of motion.

ACCELERATION: Acceleration is the rate at which velocity changes with time, in terms of both speed and direction i.e. rate of change of velocity w.r.t. time. A point or an object moving in a straight line is accelerated if it speeds up or slows down—motion in a circle is accelerated even if the speed is constant because the direction is changing continuously.

KINETIC ENERGY: The kinetic energy of an object is the energy that an object possesses due to its motion—it is defined as the work needed to accelerate a body of a given mass from rest to its stated velocity. Having gained this energy during acceleration, the body maintains this kinetic energy unless its speed changes; the same amount of work is done by the body when decelerating from its current speed to a state of rest.

POTENTIAL ENERGY: The stored energy that depends upon the relative position of various parts of a system is called potential energy. It is the energy held by an object because of its position relative to other objects and stresses within itself.

ELECTRIC CHARGE: Electric charge is the physical property of a matter that causes it to experience a force when placed in a field. A charge is the characteristic of a matter unit expressing the extent to which it has more or fewer electrons than protons.

MAGNETIC FIELD: The magnetic field is a vector field that describes the interaction between moving electric charges and magnetic materials. A moving charge in a magnetic field experiences a force perpendicular to its velocity and the magnetic field.

MAGNETIC FLUX: Magnetic flux is a measurement of the total magnetic field which passes through a given area—it is a useful tool for helping to describe the effects of the magnetic force on something occupying a given area. Magnetic flux is usually measured with a flux meter, which contains measuring coils and electronics, that evaluates the change of voltage in the measuring coils to calculate the measurement of magnetic flux. The SI unit of magnetic flux is Weber and CGS unit is Maxwell.

SPIN: Spin is the intrinsic form of angular momentum carried by elementary particles, and thus by composite particles (hadrons) and atomic nuclei. Spin is one of two types of angular momentum in quantum mechanics, the other being orbital angular momentum. (For photons, spin is the quantum-mechanical counterpart of the polarization of light; for electrons, the spin has no classical counterpart).

ELECTRIC CURRENT: An electric current is a stream of charged particles, such as electrons or ions, moving through an electrical conductor or space. It is measured as the net rate of flow of electric charge through a surface or into a control volume.

VOLTAGE: Voltage is the pressure from an electrical circuit's power source that pushes charged electrons (current) through a conducting loop, enabling them to do work such as illuminating light. In brief, it can be said that voltage is equal to pressure and is measured in volts (V).

WAVE: A wave is a propagating dynamic disturbance (change from equilibrium) of one or more quantities, sometimes described by a wave equation. Waves can be periodic; in which case those quantities oscillate repeatedly about an equilibrium (resting) value at some frequency.

STATICS: The subdivision of mechanics is concerned with the forces that act on bodies at rest under equilibrium conditions. A force acting on an object is said to be a static force if it does not change the size, position, or direction of that particular object—the force applied to a structure act as a load to that particular structure, for which static force is also known as 'static load'.

DYNAMICS: Dynamics is the branch of physical science and subdivision of mechanics that is concerned with the motion of material objects about the physical factors that affect force, mass, momentum, and energy. This branch of Physics deals with forces and their relation primarily to motion but sometimes also to equilibrium.

RADIATION: Radiation is the energy that comes from a source and travels through space and may be able to penetrate various materials. Light, radio, and microwaves are types of radiation types nonionizing.

HEAT: Heat is a form of energy that is transferred from one body to another as the result of a temperature difference—if two bodies at different temperatures are brought together, energy is transferred as heat flows from the hotter body to the colder. The direction of the energy flow is from the substance of higher temperature to the substance of lower temperature. Heat is measured by the units of energy, usually calories or joules.

TEMPERATURE: Temperature is the measure of the degree of hotness or coldness of a body expressed in terms of any of several scales, including Fahrenheit and Celsius. Temperature indicates the direction in which heat energy will flow spontaneously i.e. from a hotter body to a colder body.

VACUUM: A vacuum is a volume empty of matter, sometimes called 'free space'. In a vacuum space, there is no matter or in which the pressure is so low that any particles in the space do not affect any processes being carried on there—it is a condition well below normal atmospheric pressure and is measured in units of pressure (Pascal).

PLASMA: Plasma is one of the four fundamental states of matter. It contains a significant portion of charged particles—ions and electrons. The presence of these charged particles is what primarily sets plasma apart from the other fundamental states of matter.

CYCLE: A cycle is a sequence of changing states that, upon completion, produces a final state that is identical to the original one. One of the successions of periodically recurring events—is a complete alteration in which a phenomenon attains a maximum and minimum value returning to a final value equal to the original one. An oscillation or cycle of the alternating phenomenon is defined as a single change from up to down to up or as a change from positive to negative and again to positive.

ROTATION: Rotation is the circular movement of an object around an axis of rotation or center of rotation. A three-dimensional object may have an infinite number of rotation axes. When an object turns around an internal axis (like the Earth turns around its axis) it is called rotation. If three-dimensional objects like the earth, moon, and other planets always rotate around an imaginary line, it is a rotational axis; if the axis passes through the body's center of mass, the body is said to rotate upon itself or spin.

PHASE: In the mechanics of vibration, phase is the fraction of a period (time required to complete a full cycle) that a point completes after last passing through the reference or zero position. In electronic signalling, phase is the definition of the position of a point of time (instant) on a waveform cycle. A complete cycle is defined as 360 degrees of phase. Phase can also be an expression of relative displacement between or among waves having the same frequency.

Further Reading

(Many original books are available only in Sanskrit, Hindi, and Bengali)

Preamble

1. **The Rigveda: The Oldest Literature of the Indians,** Kagei, A. (translated by R. Arrowsmith, (1880) Ginn and Company (Boston); https://archive.org/details/dli.granth.85686
2. **Rig-Veda Sanhita; the Sacred Hymns of the Brahmans**, together with the Commentary of Sayanacharya in Six Volumes, Vol. III, Muller, Max (Ed.) (1856), William H. Allen and Co. (London); https://archive.org/details/dli.granth.85724
3. **Indian Philosophy, Vol-2,** Radhakṛishnan, S. (1980) Oxford University Press, New Delhi.
4. **Encyclopedia of religion and ethics, Vol-4,** Hastings, James. (1908) T & T Clark, Edinburgh.
5. **History of India Cosmogonical Ideas,** (1971) Bhattacharyya, Narendra Nath: Munshiram Manoharlal, New Delhi.
6. **The Power of Stars How Celestial Observations Have Shaped Civilization**, Bryan E. Penprase (2011) Springer-Verlag, New York.
7. **Revealed Mysteries of the World Divine (A Critical Evaluation of Practical Experiences of the Author in Yoga with Reference to Modern Science)** Nigurananda (Sachidananda Sarkar), KARUNA PRAKASHANI, Kolkata, India.

Chapter 1

8. **Mundaka Upanishad (first Mundaka),** S. Sankarama Sastri, (1888). https://archive.org/details/in.ernet.dli.2015.367355/page/n1/mode/2up
9. **The Markandeya Purana**, Bibek Debroy (2019) Penguin Classics.
10. **The Vishnu Purana,** Wilson, H. H. (1972) Calcutta, Punthi Pustak.
11. **A Search in Secret India,** Brunton, P. (1970) Samuel Weiser Inc., York Beach, Maine 03910. http://sageevans.businesscatalyst.com/downloads/Paul-Brunton-A-Search-in-Secret-India.pdf
12. **The Ancient Indus Urbanism, Economy, and Society**, Rita P. Wright (2009) Cambridge University Press.

Chapter 2

13. **Legends of the Jews Vol I: The First Things Created** Ginzberg, Louis (1909) (Translated by Henrietta Szold) Jewish Publication Society, Philadelphia.
14. **The Search for God in Ancient Egypt**, Jan Assmann (2001) Cornell University Press.
15. **Sumerian Mythology,** Samuel Noah Kramer (1998) University of Pennsylvania Press.
16. **COSMOS**, Carl Sagan, Random House www.goodreads.com/book/show/55030.Cosmos

17. **The Tao of Physics: An Exploration of the Parallels Between Modern Physics and Eastern Mysticism**, Fritjof Capra (1982) Flamingo, www.goodreads.com/book/show/10238.The_Tao_of_Physics
18. **Vivekachudamani of Sri Sankaracharya**, Swami Madhavananda (2019) Wentworth Press.
19. **Astavakra Samhita**, Swami Nityaswarupananda (2008) Vedanta Press & Bookshop.
20. **Sri Sri Ramakrishna Kathamrita: According to M. (Mahendra) a Son of the Lord and Desciple,** Dharm Pal Gupta (2011) Sri Ma Trust and Ramakrishna Mission.
21. **Jnana Yoga: The Yoga of Knowledge,** Swami Vivekananda (2011) Vedanta Press & Bookshop.

Chapter 3

22. **Sri Brahma-Samhita,** Bhaktisiddhanta Sarasvati, Goswami, (trans.), *with commentary by Srila Jiva Goswami*, Sri Gaudiya Math 1932, reprint The Bhaktivedanta Book Trust, Los Angeles, 1985.
23. **The Bhagavad-Gita as it is**, Prabhupada A. C. B. S., Bhaktivedanta Book Trust, USA, 1994. www.goodreads.com/book/show/1100571.Bhagavad_Gita_as_It_is
24. **Sādhan Samar: Battle of the Sacred QuestorDevī Māhātmya: Glory of the Goddess** A Spiritual Commentary on Śri Śri Caṇḍī, Brahmaṛṣi Śri Satyadeva http://sadhansamar.blogspot.com/
25. **Notes on Religion and Philosophy (1987**) by Gopinath Kaviraj (Author), Gaurinath Sastri (Editor) https://archive.org/details/QXrC_notes-on-religion-and-philosophy-gopinath-kaviraj
26. **Outgrowing God: A Beginners Guide**, Richard Dawkins, Transworld Publishers Ltd, www.goodreads.com/book/show/43532557-outgrowing-god
27. **The Physics of Theism: God, Physics, and the Philosophy of Science**, Jeffrey Koperski, Wiley-Blackwell; www.wiley.com/en-us/The+Physics+of+Theism%3A+God%2C+Physics%2C+and+the+Philosophy+of+Science-p-9781118932810

Chapter 4

28. **The Universe Speaks in Numbers: How Modern Math Reveals Nature's Deepest Secrets,** Graham Farmelo, Basic Books, www.goodreads.com/book/show/45362835-the-universe-speaks-in-numbers
29. **Sankhya Darshanam Ed. 3rd**, Bhattacharya, Panchanan Tarkaratna, https://archive.org/details/in.ernet.dli.2015.454290
30. **SĀMKHYAKARIKA (SK) of Isvarakṛsna** with the Tattvakaumudi of Vacaspati Misra, Eng. Trans by Swami Virūpakshananda, Sri Ramkṛishna, Math, Madras, 1995.
31. **Secret of Sankhya: Acme of Scientific Unification** [2013–14], Transliterated from the Sankhya Karika by Ishwara Krishna, G. Srinivasan www.goodreads.com/book/show/23111498-secret-of-sankhya
32. **Is There Anything Like Indian Logic? Anumāna, 'Inference' and Inference in the Critique of Jayarāśi Bhaṭṭa**, Piotr Balcerowicz, J Indian Philos (2019) 47:917–946; https://doi.org/10.1007/s10781-019-09400-6
33. **Vidyānandin's Discussion with the Buddhist on Svasamvedana, Pratyaksa and Pramana,** Jayandra Soni, Journal of Indian Philosophy (2019) 47:1003–1017; https://doi.org/10.1007/s10781-019-09405-1

34. **Classical Sāṁkhya on the Relationship between a Word and Its Meaning,** Olena Łucyszyna; J Indian Philos (2016) 44:303–323, DOI 10.1007/s10781-014-9264-1.
35. **The Principia: The Authoritative Translation and Guide: Mathematical Principles of Natural Philosophy,** by Sir Isaac Newton, I. Bernard Cohen, Anne Whitman, Julia Budenz, University of California Press; 1st edition (2016).
36. **The Metaphysical World of Isaac Newton: Alchemy, Prophecy, and the Search for Lost Knowledge** John Chambers, Destiny Books; 1st edition (February 13, 2018).

Chapter 5

37. **Lectures on Quantum Mechanics,** by Paul A. M. Dirac, Dover Publications; 59465th edition (2001).
38. **Mechanics: Course of Theoretical Physics, Vol. 1,** by L. D. Landau, Butterworth-Heinemann (1976).
39. **Classical Theory of Fields: Course of Theoretical Physics, Vol. 2,** by L. D. Landau, Butterworth-Heinemann (1980).
40. **Classical Mechanics,** Herbert Goldstein, Pearson (2011).
41. **Theory of Elasticity: Volume 7 (Theoretical Physics),** L. D. Landau, L. P. Pitaevskii, Butterworth-Heinemann (1986).
42. **The Feynman Lectures on Physics, Vol. I: The New Millennium Edition: Mainly Mechanics, Radiation, and Heat,** Richard P. Feynman, Robert B. Leighton, and Matthew Sands Basic Books; New Millennium edition (2011).
43. **Relativity: The Special and the General Theory—100th Anniversary Edition,** Albert Einstein, Princeton University Press (March 12, 2019).
44. **Introduction to Special Relativity,** James H. Smith, Dover Publications (2015).
45. **Introducing Einstein's Relativity: A Deeper Understanding,** by Ray d'Inverno and James Vickers, Oxford University Press; 2nd edition (2022).

Chapter 6

46. **Michael Faraday: Father of Electronics,** by Charles S. Ludwig Jr. Herald Press; Edition Unstated (1978).
47. **Quantum Electrodynamics: Course of Theoretical Physics—Vol. 4,** L. D. Landau, Butterworth-Heinemann.
48. **Electrodynamics of Continuous Media: Volume 8,** by L. D. Landau, L. P. Pitaevskii, Butterworth-Heinemann (1984).
49. **The Feynman Lectures on Physics, Vol. II: The New Millennium Edition: Mainly Electromagnetism and Matter**
50. Richard P. Feynman, Robert B. Leighton, and Matthew Sands Basic Books; New Millennium ed. edition (October 4, 2011).
51. **The Grand Design,** Hawking, S. and Mlodnow, L. (2010) Bantam Books, New York.
52. **The Universe in a Nutshell,** Stephen William Hawking (2001) Bantam.
53. **The Logic of Scientific Discovery,** Karl R. Popper (2014) Martino Fine Books.
54. **Philosophy of Science: A Contemporary Introduction,** Alex Rosenberg and Lee McIntyre (2019) Routledge.

55. **'Superlattice Structure of Disordered Carbon', S. Bhattacharyya** in CRC Concise Encyclopedia of Nanotechnology Eds. B. I. Kharisov, O. V. Kharissova, U. Ortiz-Mendez, Taylor & Francis (2015). ISBN 9781466580343.
56. **A Vortex Theory of the Electro-Magnetic Field**, Edmund Joseph Rendtorff, (2015) Andesite Press.
57. **Local Imaging of Magnetic Flux in Superconducting Thin Films: From Pinning to Depinnig of Vortices**, Tetyana Shapoval, (2011) Suedwestdeutscher Verlag fuer Hochschulschriften.
58. **'The Large-N Limit of Superconformal Field Theories and Supergravity'**, Juan Maldacena, *Int. J. Theo. Phy.* **38**, 1113 (1999).
59. **A Theory About Everything—Maldacena Closes in on One of Universe's Deepest Mysteries'** https://news.harvard.edu/gazette/story/2000/02/a-theory-about-everything-maldacena-closes-in-on-one-of-universes-deepest-mysteries/

Chapter 7

60. **What Is Life?: The Intellectual Pertinence of Erwin Schrödinger,** Hans Ulrich Gumbrecht, Stanford University Press (February 28, 2011).
61. **What Is Life?: With Mind and Matter and Autobiographical Sketches**, Erwin Schrödinger, Roger Penrose—Foreword.
62. **Space-Time Structure,** Erwin Schrödinger, Cambridge University Press; Revised edition (1985).
63. **The Principles of Quantum Mechanics,** P. A. M. Dirac, Clarendon Press; 4th edition (1982).
64. **Quantum Mechanics**, L. D. Landau and E. M. Lifshitz, Butterworth-Heinemann (1980).
65. **Quantum Mechanics—Nonrelitavistic Theory: Course of Theoretical Physics—Vol. 3**, L D Landau, Butterworth-Heinemann (1981).
66. **The Feynman Lectures on Physics, Vol. III: The New Millennium Edition: Quantum Mechanics,** Richard P. Feynman, Robert B. Leighton, and Matthew Sands Basic Books; New Millennium ed. edition (2011).

Chapter 8

67. **Introduction to Superconductivity**, Michael Tinkham (2004) Dover Publications.
68. **Superconductivity: Revised Edition Revised ed.** by V. L. Ginzburg and E. A. Andryushin (2004) WSPC.
69. **Modern Condensed Matter Physics** by S. M. Girvin and K. Yang, (2019) Cambridge University Press.
70. **'Effects of Rashba-Spin-Orbit Coupling on Superconducting Boron-Doped Nanocrystalline Diamond Films: Evidence of Interfacial Triplet Superconductivity'** S. Bhattacharyya, D. Mtsuko, C. Allen, and C. Coleman, New J. Phys. **22**, 093039 (2020) https://doi.org/10.1088/1367-2630/abafe9
71. **'Superconducting Diamond as a Platform for Quantum Technologies'** C. Coleman, F. Mazhandu, S. J. Reddhi, T. Aslan, D. Wei, C. Huynh, P. Gnauck, and S. Bhattacharyya, JPhCS **1461**, 012014 (2020). doi:10.1088/1742-6596/1461/1/012014
72. **'Finite Bias Evolution of Bosonic Insulating Phase And Zero Bias Conductance in Boron-Doped Diamond: A Charge-Kondo Effect'**, D. Mtsuko, C. Coleman, and S. Bhattacharyya, EPL **124**, 57004 (2018). arXiv:1606.06672

73. **'Enhanced Magnetic Properties and Spin Valve Effects in Gadolinium Carbon Nanotube Supramolecular Complex'** S. Ncube, C. Coleman, A. Strydom, E. Flahaut, A. de Sousa, and S. Bhattacharyya (Scientific Reports **8**, 8057 (2018).
74. **'Signatures of Two Dimensional in Superconducting Nanocrystalline Boron-Doped Diamond Films'** C. Coleman and S. Bhattacharyya, EPL, **122**, 57004S (2018). doi: 10.1209/0295-5075/122/57004
75. **'Coherent Quantum Transport Features in Carbon Superlattice Structures'** R. McIntosh, S. J. Henley, S. R. P. Silva, and S. Bhattacharyya, *Scientific* Reports **6**, 35526 (2016).
76. **'Resonant Tunneling and Fast Switching in Amorphous-Carbon Quantum-Well Structures'** S. Bhattacharyya, S. J. Henley, E. Mendoza, L. Gomez-Rojas, J. Allam, and S. R. P. Silva, Nature Materials **5**, 19 (2006).

Chapter 9

77. **Modeling of a Superconducting Sensor with Microring-Embedded Gold-Island Space–Time Control,** M. Bunruangses, A. E. Arumona, P. Youplao, N. Pornsuwancharoen, K. Ray, and P. Yupapin, Journal of Computational Electronics **19**, 1678–1684 (2020).
78. **Micro Black Hole Characteristics Using Microring Kerr Effect Space-Time Distortion Circuit,** P. Phatharacorn, A. Garhwal, M. A. Jali, P. Youplao, K. Ray, I. S. Amiri, P. Yupapin, M. A. Palomino, M. Toledo-Solano, J. Faubert, J. E. Lugo, and J. Ali, Crystals, **11**, xx (2021).
79. **Microplasma Source Circuit Using Microring Space–Time Distortion Control,** A. Garhwal, A. E. Arumona, K. Ray, P. Youplao, S. Suwandee and P. Yupapin, IEEE Transactions on Plasma Science, **48**, 3600–3605 (2020).
80. **Human-like Stereo Sensors Using Plasmonic Antenna Embedded MZI with Space–Time Modulation Control,** A. Garhwal, A. E. Arumona, P. Youplao, K. Ray, I. S. Amiri, and P. Yupapin, Chinese Optics Lett. **19**, 101301 (2021).
81. **Realizing Highly Entangled States in Asymmetrically Coupled Three NV Centers at Room Temperature,** D. Mahony and S. Bhattacharyya, Appl. Phys. Lett. **118**, 204004 (2021).
82. **Hybrid Spin-Superconducting Quantum Circuit Mediated by Deterministically Prepared Entangled Photonic States,** K. Mathieson and S. Bhattacharyya, AIP Advances **9**, 115111 (2019).
83. **Quantum Interference Correction to Coupled SYK Model: Simulation of a Traversable Wormhole,** S. Bhattacharyya and S. Bhattacharyya, (unpublished).
84. **Strange Metals and the AdS/CFT Correspondence,** S. Sachdev, J. Stat. Mech. **1011**, P11022 (2010); **Holographic Metals and the Fractionalized Fermi Liquid,** S. Sachdev, Phys. Rev. Lett. **105**, 151602 (2010).
85. **Sachdev-Ye-Kitaev Non-Fermi-Liquid Correlations in Nanoscopic Quantum Transport,** A. Altland, D. Bagrets, and A. Kamenev, Phys. Rev. Lett. **123**, 226801 (2019).
86. **Remarks on the Sachdev-Ye-Kitaev Model,** J. Maldacena and D. Stanford, Phys. Rev. D **94**, 106002 (2016).
87. **Building a Holographic Superconductor,** S. A. Hartnoll, C. P. Herzog, and G. T. Horowitz, Phys. Rev. Lett. **101**, 031601 (2008).
88. **The Soft Mode in the Sachdev-Ye-Kitaev Model and its Gravity Dual.** A. Kitaev and S. J. Suh, *J. High Energ. Phys.* **2018**, 183 (2018).
89. Eternal Traversable Wormhole, J. Maldacena and X. L. Qi, arXiv:1804.00491v3; **Diving into traversable wormholes,** J. Maldacena, D. Stanford, and Z. Yang, *Fortsch. Phys.* **65**, 1700034 (2017).
90. The Coupled SYK Model at Finite Temperature, X. L. Qi and P. Zhang, JHEP **05**, 12 (2020).

91. **Creating and Probing the Sachdev–Ye–Kitaev Model with Ultracold Gases: Towards Experimental Studies of Quantum Gravity,** I. Danshita, M. Hanada, M. Tezuka, Prog. Theor. Exp. Phys. **2017**, 083I01 (2017); **How to Make a Quantum Black Hole with Ultra-Cold Gases,** I. Danshita, M. Hanada, M. Tezuka, arXiv: 1709.07189v1.
92. **Quantum Simulation of the Non-Fermi-Liquid State of Sachdev-Ye-Kitaev Model,** Z. Luo, Y. Z. You, J. Li, C. M. Jian, D. Lu, C. Xu, B. Zeng, and R. Laflamme, npj Quantum Information **5**:53 (2019).
93. **Mimicking Black Hole Event Horizons in Atomic and Solid-State Systems,** M. Franz and M. Rozali, Nature Rev. Mater. **3**, 491–501 (2018).
94. **Quantum Holography in a Graphene Flake with an Irregular Boundary,** A. Chen, R. Ilan, F. de Juan, D. I. Pikulin, and M. Franz, Phys. Rev. Lett. **121**, 036403 (2018).
95. **Simulation Model for Complexity in Black Holes and Demonstration of Power of One Clean Qubit Using IBM QX,** Dharmaraj, B. K. Behera, and P. K. Panigrahi, Quantum Stud.: Math Found. **8**, 167–178 (2021).
96. **Traversable Wormhole Dynamics on a Quantum Processor**, D. Jafferis, A. Zlokapa, J. D. Lykken, D. K. Kolchmeyer, S. I. Davis, N. Lauk, H. Neven, and M. Spiropulu, *Nature*, **612**, 51–55 (2022) https://doi.org/10.1038/s41586-022-05424-3.

Chapter 10

97. **Physics of the Future: How Science Will Shape Human Destiny and Our Daily Lives by the Year 2100**, *M. Kaku*, Anchor, https://physicstoday.scitation.org/doi/abs/10.1063/PT.3.1299?journalCode=pto
98. ***Quantum Enigma: Physics Encounters Consciousness***, Bruce Rosenblum and Fred Kuttner, Oxford University Press; www.goodreads.com/book/show/100027.Quantum_Enigma *see also* www.researchgate.net/publication/28765349_Review_of_the_Book_Quantum_Enigma_-_Physics_Encounters_Consciousness
99. **Consciousness in the Universe: Neuroscience, Quantum Space-Time Geometry and Orch OR Theory,** R. Penrose R, and S. Hameroff, Journal of Cosmology, Vol. 14., JournalofCosmology.com (2011).
100. **Tricking the Uncertainty Principle**, Caltech, 05 May 2014, www.caltech.edu/news/tricking-uncertainty-principle-42816
101. **The Turning Point: Science, Society, and the Rising Culture,** Fritjof Capra, Bantam, https://physicstoday.scitation.org/doi/10.1063/1.2914857
102. **Quantum Computation and Quantum Information,** Michael A. Nielsen and Isaac L. Chuang Cambridge University Press (2011).
103. **Learn Quantum Computing with Python and IBM Quantum Experience: A hands-on introduction to quantum computing and writing your own quantum programs with Python**, Robert Loredo Packt Publishing (2020).
104. **Artificial Intelligence: A Modern Approach**, Stuart Russell and Peter Norvig Pearson (2020).
105. **Building Computer Vision Applications Using Artificial Neural Networks: With Step-by-Step Examples in OpenCV and TensorFlow with Python**, Shamshad Ansari Apress (2020).
106. **'Realizing Highly Entangled States in Asymmetrically Coupled Three NV Centers at Room Temperature'** D. Mahony and S. Bhattacharyya, Appl. Phys. Lett. **118**, 204004 (2021) DOI: 10.1063/5.0043334

107. **'Demonstrating Geometric Phase Acquisition in Multi-Path Tunnel Systems Using a Near-Term Quantum Computer'**, Shaman Bhattacharyya and S. Bhattacharyya, J. Appl. Phys. **130**, 034901 (2021). https://doi.org/10.1063/5.0049728
108. **'Experimental Simulation of Hybrid Quantum Systems and Entanglement on a Quantum Computer'** F. Mazhandu, K. Mathieson, C. Coleman, and S. Bhattacharyya, Appl. Phys. Lett. **115**, 233501 (2019).
109. **'Holonomic Control of a Three-Qubits System in an NV Center Using a Near-Term Quantum Computer'**, S. Bhattacharyya and S. Bhattacharyya, Entropy, **22**, 1593 (2022).

Index